Urban Agriculture

The *Urban Agriculture* Book Series at Springer is for researchers, professionals, policy-makers and practitioners working on agriculture in and near urban areas. Urban agriculture (UA) can serve as a multifunctional resource for resilient food systems and socio-culturally, economically and ecologically sustainable cities.

For the Book Series Editors, the main objective of this series is to mobilize and enhance capacities to share UA experiences and research results, compare methodologies and tools, identify technological obstacles, and adapt solutions. By diffusing this knowledge, the aim is to contribute to building the capacity of policy-makers, professionals and practitioners in governments, international agencies, civil society, the private sector as well as academia, to effectively incorporate UA in their field of interests. It is also to constitute a global research community to debate the lessons from UA initiatives, to compare approaches, and to supply tools for aiding in the conception and evaluation of various strategies of UA development.

The concerned scientific field of this series is large because UA combines agricultural issues with those related to city management and development. Thus this interdisciplinary Book Series brings together environmental sciences, agronomy, urban and regional planning, architecture, landscape design, economics, social sciences, soil sciences, public health and nutrition, recognizing UA's contribution to meeting society's basic needs, feeding people, structuring the cities while shaping their development. All these scientific fields are of interest for this Book Series. Books in this Series will analyze UA research and actions; program implementation, urban policies, technological innovations, social and economic development, management of resources (soil/land, water, wastes…) for or by urban agriculture, are all pertinent here.

This Book Series includes a mix of edited, coauthored, and single-authored books. These books could be based on research programs, conference papers, or other collective efforts, as well as completed theses or entirely new manuscripts.

Silvio Caputo

Small Scale Soil-less Urban Agriculture in Europe

Springer

Silvio Caputo
Kent School of Architecture and Planning
University of Kent
Canterbury, UK

ISSN 2197-1730 ISSN 2197-1749 (electronic)
Urban Agriculture
ISBN 978-3-030-99961-2 ISBN 978-3-030-99962-9 (eBook)
https://doi.org/10.1007/978-3-030-99962-9

This Springer imprint is published by the registered company Springer Nature Switzerland AG
The registered company address is: Gewerbestrasse 11, 6330 Cham, Switzerland

Contents

List of Figures

List of Tables

Chapter 1
Introduction

Abstract This chapter sets the context and the rationale for this book's investigation into small scale soil-less agriculture projects in European cities. Macro factors such as demographic expansion, an estimated need for increased food production, and the high negative environmental impact of conventional agriculture and globalised food supply constitute formidable challenges that require bold answers. Urban agriculture and technology are increasingly presented as solutions for such challenges. Soil-less food production technologies are water and space efficient, thus being particularly suitable for an urban environment. Small to medium scale urban agriculture projects based on these technologies can be seen across Europe, although the overall urban soil-less sector contribution to meet global food demand is still limited.

Urban agriculture is practiced mainly at a small scale, with urban farmers typically driven by social and environmental values, often prioritising issues such as food security, wellbeing, and the environment over strictly commercial interests. The rationale for investigating small scale soil-less urban agriculture in Europe is that, although still in its infancy, this phenomenon is growing. As such, its impact on those values that characterise urban agriculture as a practice generating multiple benefits, concurring to strengthen urban resilience, must be questioned. Are soil-less technologies changing the perception of what is natural and sustainable? Are they changing the relationship between urban farmers and technology at large? Are they reframing urban farmers' social and environmental goals? The chapter describes the context within which these questions are formulated and ends with a brief outline of the structure of the book.

The purpose of this book is to investigate the reasons why urban farmers choose soil-less technologies to grow food. It documents how and discusses why such technologies, which are typically used in industrial high-tech food production with the deployment of sophisticated and costly equipment, are now also being used in a completely different context: in cities, at a small scale and with equipment that is affordable because it can be self-assembled. This scope may seem very specific and perhaps unusual, especially considering that, when compared to the extent to which urban agriculture is practiced either at a household, or at a community, or at a

S. Caputo, *Small Scale Soil-less Urban Agriculture in Europe*, Urban Agriculture, https://doi.org/10.1007/978-3-030-99962-9_1

commercial level, small scale soil-less projects represent a niche phenomenon. So, why write about soil-less technologies applied to urban agriculture, when this is not yet a consolidated trend? This introductory chapter will outline the motivations for this book and detail some of the topics that will be elaborated on over the following chapters. Such topics are well known to those familiar with the ongoing urban agriculture debate. They include the material and immaterial benefits that urban agriculture can generate, although such benefits are revisited within a perspective of soil-less production. Factors such as the relationships between technology and people, and technology and nature, which are quite new to the debate, are discussed.

It is necessary to clarify the use of some recurrent terms at the onset of this book. *Urban agriculture* is practiced in many ways and the terms characterising people practicing it vary accordingly. *Urban gardeners*, for example, can be associated with growing food and ornamental plants for leisure, in allotments. At the other end of the spectrum, *urban farmers* can be associated with those who practice urban agriculture professionally. The case studies presented in the book are either community projects or small enterprises led by people who may have no horticulture or farming professional training but are working full-time to produce food, hence the choice to call them farmers. *Soil-less* is another recurrent word and refers to technologies such as hydroponics and aquaponics, which are used to grow food in absence of soil. Mushroom farms, included in the case studies, use substrate although this is often not real soil, but rather a substrate designed for this cultivation. *Controlled environment agriculture* is a term that could encompass all these food production technologies, although, in the case studies presented here, only a few are equipped with indoor environmental control systems. In this book, the term *soil-less* is used for hydroponic, aquaponic and mushroom production to capture either the absence of soil or an indoor cultivation that uses substrates alternative to natural soil. There are other technologies of indoor food production requiring no soil and that can be practiced in urban environments such as insect or algae farming. The scope of the book, however, is restricted to these three technologies because they are already established and trialled. *Hydroponic* and *aquaponic* technologies are still evolving but, like mushroom farms, they have supply chains and technology providers both for high-tech and low-tech applications, which allow community groups or small enterprises to utilise them. Other types of new food production (e.g., aeroponics and insect farming) have not reached that stage yet.

In arguing for the relevance of this book, it is useful to frame the topics debated within the wider context of food security and to bring to mind a few well known, but nonetheless alarming, facts. Projections of demographic growth suggest that the global population will reach 9 billion by 2050 (United Nations, 2004). Consequently, the Food and Agriculture Organisation of the United Nations (FAO) estimate an increase in food demand of approximately 60% (FAO, 2011). It is difficult to imagine how food production can increase without further damaging the planet's supporting systems, which industrial agriculture has already compromised in terms of resource availability, biodiversity loss, carbon sink capacity (Bruinsma, 2003) and human health generally (Horrigan et al., 2002). Agricultural cultivation occupies about 12% of the total land surface on the planet (FAO, 2015) and is the highest emitter of Nitrous oxide (N_2O), an important anthropogenic greenhouse gas (Reay et al., 2012). A study by Hedenus et al. (2014), which looks at ways to lower GHG emissions connected to food production and keep temperature rise below 2°, shows that farming intensification can reduce emissions from this sector if well managed. But in light of a demographic expansion and increase in demand, only dietary changes – therefore drastic reduction of meat consumption – can reduce GHG emissions. Decreasing food waste could also contribute to avoid critical temperature rise. Globally, one quarter to one third of the food produced for human consumption is lost or wasted, with associated emissions being approximately 2.5Gt (Guo et al., 2020). A 2014 FAO report quantifies land and water used to grow food that is wasted, equivalent to 0.9M ha and 49 Gt of CO_2 (Lemaire & Limbourg, 2019). Furthermore, reduction in meat and dairy consumption could also help attain climate goals. In 2017, emissions from the livestock sector in the 28 EU countries were 81–86% of the total emissions from the agricultural sector (10% of EU GHG emissions) (Peyraud & MacLeod, 2020). However, these estimates do not include GHG generated outside the EU from the production of animal feed and fertiliser. Bellarby et al. (2013) estimated that in Europe, 2007 GHG emissions of all livestock products amounted to 12–17% of total EU emissions, once emissions connected to land use and land use change were included.

The current agricultural system is water intensive, being responsible for 85% of global water consumption (Shiklomanov & Rodda, 2003). Irrigation approaches can also contribute to unnecessary water use. In a study by Mueller et al. (2012), water usage for irrigation is assessed globally in order to analyse geographical areas that could increase their yield with more water and others that are too profligate. Yields of staple crops such as wheat, maize and rice could be increased with a correct dosage of fertiliser and water, depending on the conditions of the soil and the local climate, and without increasing current overall global use of such resources. Intensification of water usage in regions currently under deficit of irrigation can increase their yield by up to 30% (Pfister et al., 2011). These studies, however, recognise that climate change is an element of high uncertainty, which will impact the validity of predictions. More importantly, there are signs that progressive increase in yields has reached a peak and, indeed, productivity is decreasing due to reduced soil fertility. In fact, after initial increments of nitrogen fertiliser, its efficiency decreases with increased application (Tilman et al., 2002). For example, over the last 15–20 years rice yields in Japan, Korea and China have not increased (ibid). Soil fertility can be increased with higher use of organic methods relying on organic matter to enhance soil quality and structure (Watson et al., 2002). However, only 8.5% of the EU agricultural land is farmed organically (Eurostat, 2021), and, as of 2016, about 1.2% globally (Willer et al., 2018).

These facts only give an approximate picture of the complexity and scale of the challenges connected to feeding the world population and yet avoid environmental collapse. In fact, in order to understand the role that soil-less technologies can play not only in terms of food production but also within the broader food system and the environment it is important to put agricultural production into a broader context which looks at the connections between demand and supply, land use changes, dietary norms, policy and environmental degradation. Big ideas are necessary that can contribute to the attainment of a sustainable food production and supply chain.

Technology is playing a major role in developing solutions for a more sustainable food production, but these can be highly contentious. Genetically modified crops, for example, some of which have been designed to be drought-resistant, pose major ethical challenges and do not address key issues of soil fertility and land scarcity. Conversely, soil-less technologies are water and space efficient, thus potentially translating into reduced demand for agricultural land. Yet, these technologies can be contentious too: they do not increase the overall biodiversity or necessarily provide some ecosystem services in urban and rural areas. More importantly, the opportunity for intensive soil-less production to release pressure on agriculture land and repurpose marginal farmland areas to increase biodiversity and regenerate soil is far from being considered in any policy debate. Hydroponics (the cultivation of plants in water) is a technology initially tested in the 1930s, building on a long history of experiments aimed at understanding the plants' physiology, and testing hypotheses about the composition of the plants' biomass and the quantity of water constituting it (Jensen, 1997a, b). Today, hydroponic technology is exploited at an industrial scale although studies show that this type of production can be more energy intensive than conventional agriculture (Baddadi et al., 2019; Barbosa et al., 2015). Another bold approach that can contribute to a more sustainable food system is urban agriculture. Food grown in cities will perhaps not feed the planet but, if effectively supported by policy, can help attaining this target not only in terms of food availability but also fair access to it (Mayer & Paul, 2021; Zezza & Tasciotti, 2010) and education leading to more sustainable diets. Assessments on the potential productivity of urban agriculture vary greatly. Ackerman et al. (2014) estimated that New York City's extended metropolitan area can support between 58 and 89 % of the city population's demand for fresh produce. A study about the urban agriculture production capacity in the same city estimated that vegetable production in existing community gardens would feed approximately 1700 people per year, but if all available urban vacant lots and other open spaces were used for food production, the amount grown would meet a vegetable intake sufficient for 55 million people (Hara et al., 2018). Nadal et al. (2017), suggested that the suitable rooftops in Rubi, Barcelona, if equipped with greenhouses, could produce 50% of the city's expected demand for tomatoes. Although these studies may be overly optimistic and do not take into account spatial and socio-economic constraints, they show that there is interest – at least from researchers - in developing strategies to increase levels of food production in cities and soil-less technologies are considered one of such strategies (Despommier, 2010; Al-Chalabi, 2015). As noted, in European cities, the uptake of soil-less technologies is still marginal and does not always have the potential to grow food in large quantities, as the case studies included in this book will show. This brings us back to the original question: why write about soil-less methods applied to urban agriculture, when this is not yet a consolidated approach?

The thesis of this book is that such a phenomenon is small but significant. Regardless of its scale, it could lead to a shift in the way urban agriculture is practiced, opening the potential for indoor spaces and non-green open areas to be repurposed for food cultivation, with some conceptual implications attached. For example, one of the motivations to grow food in cities, at least from a Global North perspective, is to enjoy nature within the built environment and to be closer to natural cycles. In a study on allotment sites in Stockholm, Barthel et al. (2010) concluded that one of the benefits of growing food in these places, with many individual gardeners sharing knowledge and experience, is the perpetuation of an ecological memory (a collective cultural tradition on the relationship between man and nature), which, in the urbanised age, is at risk of disappearing. By exchanging information on techniques for growing, and on pest management and fertilisation, gardeners ensure that an accumulated knowledge – which is deeply connected to the natural environment – is preserved. One of the results of the process of urbanisation is the progressive detachment of citizens from the surrounding natural environment and the life-supporting systems that it provides (Giusti et al., 2014; Andersson et al., 2014). This detachment can be accelerated with the loss of natural habitat in cities due, for example, to land use changes in the peri-urban landscapes of cities, often rich in biodiversity (Turner et al., 2004). Urban and peri-urban green areas, including food growing spaces, can therefore greatly contribute to strengthen our respect and understanding of nature (Giusti et al., 2014) through the provision of ecosystems cultural services (Raymond et al., 2018), whilst enriching biodiversity (Gren & Andersson, 2018). However, urban farmers engaging with soil-less technologies enjoy a form of nature that grows without soil, sometimes indoors and disconnected by seasonal weather cycles. Can this engender long-term consequences for the way urban farmers understand nature and its enjoyment? Is this potentially changing the relationship between urban and rural dimensions? Is this diminishing, in the farmers' perception, the importance of protecting and augmenting urban green infrastructure?

Another key motivation to grow food in cities is the provision of healthy food. This motivation is connected with the broader issue of food security and access to quality food for low-income groups. The role that urban agriculture can play in securing food to those on the poverty line is fully recognised for developing countries (see for example Badami & Ramankutty, 2015; Adeoti et al., 2012; Karanja et al., 2010), but much less in the context of mature economies in which social support for low-income groups is usually available and income disparities less extreme. Yet, over the last two decades, food poverty has come to the forefront dramatically in Europe and, perhaps more significantly, in the US (Morgan, 2015). As of 2014, up to 14% of the population of the US and 9% of the EU27 population have experienced food insecurity (Borch & Kjærnes, 2016). Links are becoming evident between low-income groups and obesity, which can manifest in excessive consumption of cheap and highly processed food such as fast food (Fraser & Edwards, 2010), and in a phenomenon such as food deserts (Burgoine et al., 2017). Policies exacerbating income gaps and ecological pressures create extreme tensions, especially in cities (Maye, 2019). Increasingly, in response to these trends, social movements demand fair access to food (De Amorim et al., 2019) also through urban agriculture, which can provide healthy food to those who cannot afford it.

The issue of access to healthy food is not only confined to low-income groups and can be culturally influenced. Food scares (i.e., public concern about contamination or shortage of foodstuff) have recently greatly contributed to increasing the mistrust of all consumers towards industrial agriculture and farming. Lack of traceability and unclear provenance of food are amongst the causes behind such food scares, sometimes only perceived, but without a real health threat (Whitworth et al., 2017). A case in point is the 2013 horse meat scandal in the UK (i.e., the presence of horse meat in foodstuff claimed to be made with different types of meat), which was not posing health risks but resulted in lack of credibility, amplified by a 'cultural' stigma of British consumers towards horse meat, which is not regarded as suitable for consumption. More generally, the negative impact of industrial agricultural practices on the environment, massively relying on pesticides and synthetic fertilisers, have been increasingly acknowledged by consumers, with organic produce increasing its market share (Saba & Messina, 2003). Also, the socio-economic profile of urban farmers can be predominantly composed of middle-class groups (Hoover, 2013), who do not grow food for subsistence but rather to ensure quality and provenance. However, urban soil can be contaminated (Sharma et al., 2015) making food grown in such soils unsuitable for consumption. Studies suggest that contaminated soil can be found more frequently in low-income neighbourhoods, thus posing issues of environmental justice (Sharma et al., 2015; McClintock, 2012). Generally, through organic techniques and organic soil enhancement, urban agriculture practices can improve soil conditions (Brown et al., 2016).

Issues related to horticultural skills necessary to grow food in cities and food poverty need to be reframed when soil-less technologies are considered. Growing food on soil requires basic tools, a plot of land and some knowledge of horticultural techniques, which is easy to access. Allotment holders can rely on the advice of more experienced gardeners within allotment sites; manuals on gardening are quite common and, although the process of urbanisation constantly threatens our collective ecological memory, many still retain it in some form. Conversely, soil-less cultivation is not part of our cultural background and only few have prior knowledge that is relevant. Soil-less cultivation requires equipment, which is not as commonly available as gardening tools, and it necessitates specialist knowledge about plant physiology, water acidity and more. It therefore requires determination in locating and acquiring this knowledge, which will not be found amongst gardeners in allotment sites. Nonetheless, despite the uncommon and apparently complex techniques necessary to manage soil-less cultivation, hydroponics and aquaponics have been used in simplified forms by many non-experts, also to address food poverty. For example, many projects have facilitated the construction of soil-less units and the training of soil-less farmers in order to provide subsistence food within low-income communities in South America (Orsini et al., 2013; Fecondini et al., 2010; Tabares, 2003). Simplified hydroponic systems are based on the principle that plants can be grown not on soil but in other substrates (such as coconut fibre) or simply in water as a growing medium and with basic equipment. FAO has extensively documented and promoted traditional forms of aquaponics such as those practiced in South-East Asian countries (e.g., India, Vietnam and China), including culture of fish in rice fields, in which fish enrich the water of paddy fields with nutrients suitable for plants. At the same time FAO has also promoted self-build, more contemporary forms of aquaponic systems, some of which have been constructed in places such as the Gaza strip and some African countries such as Ethiopia, Ghana and Nigeria, in order to alleviate food scarcity and for humanitarian relief (Somerville et al., 2014; Halwart & Gupta, 2004; FAO, 2011). As one of the case studies of this book documents, subsistence small-scale aquaponics at a household level, are no longer promoted only in conflict zones and developing countries, and have been trialled in Spain too, in a very poor neighbourhood in Seville.

The attitude of science – as well as that of consumers – towards alternative options to mainstream agricultural production has changed over the course of the past two decades. At the beginning of the century, it was possible to read in Nature, one of the most prestigious scientific journals, commentaries debunking the advantages of organic production as a myth, refuting that pesticides have a long-term impact on insect populations and stating that only transient negative effects from their use are reported (Trewavas, 2001). Today, people's perception of organic food is largely associated with a higher quality when compared to conventional food. In fact, organic food sales have been constantly growing over the last decade. In the EU, organic cropland increased 89% in the period 2004–2015, reaching 5,967,854 ha. Sales of organic produce grew from €15.9 billion in 2008 to €27.1 billion in 2015 (Bostan et al., 2019). Against the value of the EU agricultural market in 2018 totalling €176.9b, the organic share is approximately 15% (Eurostat, 2019). European customers purchasing organic food, in particular those who regularly purchase it, are driven primarily by health and safety issues as well as – particularly in Northern Europe – animal welfare (Naspetti & Zanoli, 2009). Absence of chemicals in the production process and additives in processed food are key factors for these customers.

Urban farms often follow organic farming methods although there is some evidence that there is an excessive and sometimes noxious use of organic matter to enhance soil, resulting in excess of elements such as Potassium, Phosphorus and Nitrogen (Wielemaker et al., 2019). Although accurate information and sometimes training is required in order to avoid an incorrect application of organic practices, a shared belief that organic is better between urban farmers is evident and is in line with the desire to live in closer connection with, and respectful of, urban nature. Moreover, studies suggest that young consumers associate urban farming with organic, even when this is not the case (Grebitus et al., 2017). Soil-less methods, however, cannot generally be defined as organic also because hydroponic – and often aquaponic – systems use synthetic nutrients to feed plants generally, although organic nutrients are being currently experimented (Phibunwatthanawong & Riddech, 2019). Furthermore, in Europe, there are several organic certifications for in-soil produce but no certification has yet been developed for soil-less products. The urban farmers interviewed for this book consider their products healthy, mainly because no pesticides are used; the cultivation of crops in a greenhouse or in any other type of controlled environment reduces the risk of pests and therefore the need for chemical pest control systems. Moreover, synthetic (or organic) fertilisers are not dispersed in the environment because water with nutrients recirculates within the soil-less systems without being discharged. Nonetheless, the acceptance and use of synthetic nutrients for crop production is a major shift from the values that are commonly held by urban farmers, which is one of the issues investigated in this book. Has this more 'flexible' approach to an understanding of organic, healthy food changed the value systems of many urban farmers? Studies suggest that, from the point of view of consumers, there is diffidence towards the use of chemicals in soil-less production and that soil-less produce is considered non-natural because of the inherent artificiality with which the food is produced (Caputo et al., 2020; Specht & Sanyé-Mengual, 2017; Food Ethics Council, 2015). Caputo et al. (2020) also provide some evidence that urban farmers who are experienced in horticultural techniques, have prior knowledge of soil-less methods of production and hold the conservation of nature as a key value, can be even more sceptical towards these techniques than non-expert growers or laypeople. Nonetheless, the current interest in soil-less suggests that there are urban farmers with different views.

In fact, one of the concepts behind this book, illustrated in Chap. 2, is that urban agriculture is in evolution and, as any phenomenon which is changing fast, is a fertile ground for experimenting with innovative arrangements and models. From a practice predominately confined to households and allotment sites, it has become a battleground for social movements claiming the Right to the City and the use of abandoned public space to grow food (e.g., the movement of guerrilla gardening and Transition Towns); a space for groups to engage with the broader community and sometimes to provide educational and social services (e.g., workshops with students and nature-based therapeutic assistance to improve mental well-being); or a key player in reframing urban food policies (Moragues-Faus & Morgan, 2015). Over the last three decades, the increased focus of research activities on urban agriculture is notable (Viljoen & Bohn, 2014). Much of this research is concerned with the identification of future trends of development. For example, Specht et al. (2014) give an account of current attempts to integrate urban agriculture with buildings, implicitly expanding the idea of continuous productive urban landscapes coined by Viljoen et al. (2012 – e.g., an approach to establish a network of food producing green areas in cities), now occupying not only open spaces but also buildings. To this end, Thomaier et al. (2015) propose the term *Zero-Acreage Farming* ('non-use of land or acreage for farming activities'). Another defining characteristic of contemporary urban agriculture is that it sometimes occupies land on a temporary basis, utilising sites that are available for a few years, until – for example – planning consent for development is granted. This leads to the creation of a mutable geography of food growing spaces, that changes in response to urban development dynamics, although the experience of the groups who manage projects that disappear or transform, outlives the project themselves (Caputo et al., 2016).

This expansion of food growing from its 'natural' place', on green areas and in-soil, to potentially everywhere, inevitably involves technology, which, this book claims, is one of the latest factors with which urban farmers engage. From green roofs to anaerobic digestion to hydroponics, technology is becoming a necessary factor, enabling a greater range of options for urban farmers and, inevitably, influencing modalities of practice, shaping the built environment and modifying behaviour. Supported by evidence on the benefits of green roofs in terms of energy conservation, mitigation of water runoffs and biodiversity (Berardi et al., 2014), cities have implemented policies that can potentially change their cityscape. Green roofs for food production could generate additional environmental benefits compared to traditional green roofs by using, for example, by-products from urban organic waste for substrates (Grard et al., 2020). Vienna, Munich and Copenhagen are some of the European cities that, through incentives, greatly expanded their total green roof surface area (Brudermann & Sangkakool, 2017). The construction of projects such as Brooklyn Grange, one of the largest urban agriculture rooftop projects in New York City, largely covered by media, is another case in point showing how technology is shaping our cities and the perception of things.

The study of the interaction of technology and people offers frames of interpretation that are useful for this book. A large part of the knowledge today comes from science (Ropohl, 1999) and, undeniably, our society is imbued with technology and the science that underpins it. Ropohl (1999) resorts to complex systems theory in order to illustrate how technology and people interact. Complex systems theory explains phenomena by viewing the elements involved in them as parts connected by relations. As parts perform actions in response to, for example, external perturbances or specific purposes they intend to attain, the relations between them trigger change in the system as a whole. Society is composed of people performing actions finalised to an outcome. In order to perform these actions, people utilise technological tools, that is, tools produced through advanced manufacturing processes. Following the complex systems logic, the nature of each action will be inevitably influenced by the tools used and their functionality. As people work by interacting with other people and within a structured context (e.g., offices, schools, factories and farms), the functionality of these technological tools will influence social relations and also the context in which they work. The internet enables home working and therefore people can work without sharing the same office space, with reduced opportunities for social interaction. In turn, demand for office space will be reduced or such a space adapted in order to be rationally utilised. Technological tools have another property, they provide outputs with no need for the user to know the process which led to their development. With pocket calculators we can obtain the square root of a number without knowing the mathematical rules to calculate it. We use computers without understanding codes enabling their functioning or without knowing how a processor is built. Nonetheless the opportunity that pocket calculators offers to perform fast mathematical operation can become embedded in daily actions and people may soon forget the mathematical operations to compute a square root. When this happens, technical developments become 'institutionalised'; they 'channel and shape the behaviour of the individuals and integrate them into a common culture' (Ropohl, 1999). Inevitably, as technology shapes culture and behaviours, these too have an impact on the way technology itself is used and transformed, in a co-evolutionary process. Users of specific technologies may be clustered in varied groups, which interact with each other, forming networks of users influencing each other (in Chap. 7, a study based on internet users forming groups of interest in soil-less technologies will elaborate on this point). A case in point is blogs in which solutions to technical glitches of computer software and hardware are offered, together with opinions and ratings on software apps or digital tools. Within this dynamic environment, innovation can happen whenever niche phenomena proposing alternative technologies (e.g., start-ups) create their own space, in which they deviate from the mainstream utilisation and understanding of such technologies. The uptake of innovative, niche use of technology happens when mainstream technology is no longer in line with evolving institutional or social values (Geels, 2004). Innovative technologies to grow and process food have changed our diets and oriented our choices in purchasing and preparing food, not always for the better. These technologies are generally deployed at an industrial scale and rely on large retail outlets. Innovation that comes from below, such as the one experimented in urban agriculture models (e.g., CSA), generates an impact on society that is sometimes competing with that triggered by large scale, market-led innovation. This is another point that the book will try to address: in what way a technology that is used for industrial food production can impact differently when deployed at a small scale, appropriated and transformed.

Although soil-less techniques to grow food have a long history, as documented in Chap. 4, technologies allowing the exploitation of these techniques at an industrial scale are relatively new. Hydroponics, for example, were designed and trialled before and during World War 2 as an alternative to industrial agriculture requiring fewer inputs to attain maximised production. Growing crops indoors, fed with synthetic nutrients, had the advantages of a higher control on pests and elimination of soil fertility enhancement, thus avoiding the use of insecticides and organic or chemical fertilisers. The experimentation with these technologies remained a niche phenomenon for a long time, until an increased demand for out-of-season vegetables and countries short of land, such as the Netherlands, needing to increase dramatically their productivity per land area, propelled it to a technology that, for example in 2018, provided 30% of the global market for tomatoes. This is the largest hydroponic crop globally produced, 'owing to its faster cultivation rate and as it requires less water compared to regular farmed tomatoes' (Grand View Research, 2020a, b). Aquaponics, which followed a trajectory of technological development similar to hydroponics, is not as diffused yet, possibly as a consequence of the inclusion of fish farming, which is complex and at a higher risk of failure. However, as noted, it is strongly promoted by FAO as an effective subsistence technology and is identified by the EU as one of the most promising technologies over the short term (Van Woensel & Archer, 2015). Mushrooms, which are also included in this book, constitute a more established technique of cultivation. Similar to hydroponics and aquaponics, highly sophisticated technologies applied to indoor mushroom farming are deployed in large scale farms. But low-tech mushroom farming is possible too, and small scale community projects based on this technique can be found in Europe.

The surge of soil-less production globally (particularly in the mushroom and hydroponic industry as documented in Chap. 4) only demonstrates how, with increasing pressure for greater rationalisation of resource use and the environmental impact of industrial agriculture, niche food production technologies are now expanding. By the same token, the utilisation of technologies for small community-led and commercial sustainable enterprises in cities, can break new ground by testing new approaches and models. The case studies presented in this book as well as a pilot study on the diffusion of soil-less technologies on the internet and YouTube in Chaps. 6 and 7, show that there is a continuous attempt to experiment with existing technology by making them more affordable, assembling them with DIY techniques. For this to happen, however, a transformation and alignment of the values motivating urban farmers in their practices, and citizens in their perception of food produced in cities, with those underpinning soil-less technologies is necessary. Is this happening? How can urban farmers and many consumers embracing organic food accept the (at least perceived) artificiality of soil-less technologies? Small urban agriculture projects are also the place where social innovation happens. Social innovation can be defined as a process of re-combining in order to attain specific aims in a new way (Manzini, 2014). Aims underpinning urban growing practices are multiple and urban growing places are often laboratories where new models are experimented with. Maye (2019) warns against the attempts to quantify outputs of urban agriculture since they may lead to a narrow view of the overall outputs, which actually include the development of community capacity. He advocates for urban agriculture to deliver radical innovation by merging technological and social innovation.

The book focuses on small scale urban soil-less practices; therefore, clarification for the choice of this scale is in order. At a very general level, this choice can be justified with the argument that urban agriculture is predominantly practiced at a small scale due to a number of constraints. Large plots of green land for cultivation are generally unavailable in cities. Besides, high land values drive dynamics that facilitate the use of land for development rather than other land uses. Also, the population of urban farmers is largely composed of people who practice for leisure and/or use the practice as an opportunity for strengthening community bonds, an objective which does not require large plots of land. Conversely, the management of large plots would require organisational structures and investments that are beyond the goals and capabilities of individuals and groups that work in this sector. In peri-urban areas, there are commercial farms that grow for a local market, on relatively large plots of land. These farms are generally fewer in number when compared to allotments and community gardens and therefore covering a limited share of the overall urban space used for cultivation. Data on the surface area of peri-urban large plots of land cultivated are not available, although a few studies have tried to measure the scale of this phenomenon using aerial mapping such as Google Earth. A study in Rome by Pulighe and Lupia (2016) for example, suggests that, as of 2013, large plots of land (termed urban farms in the study, on the basis of the scale of the activity), occupy 9.8% of the total land used for urban cultivation. Allotments (termed community gardens), institutional gardens (e.g., school gardens) and illegal gardens total 26.7%, with the remaining land occupied by residential, private gardens where food is grown. The share of land occupied by farms changes according to each city's morphological, political and socio-cultural context. A study in Chicago suggests that the real share in surface area of farms (as opposed to data available on registered membership lists of associations representing the sector) is only 4.7% as opposed to 22.3% for allotments and school gardens (Taylor & Lovell, 2012). In both estimates, the share of small plots prevails, and this is likely to happen in other cities in the global north. This seems to support the assumption that urban agriculture is mainly practiced at a small a scale and, because of the way cities are organised, it is likely to retain this characteristic. The focus on the small scale enables therefore a study of the sector through actors and practices that are more representative of the reality on the ground.

The social dimension of urban agriculture can manifest itself also in commercial projects. In a study on USA urban farms, Dimitri et al. (2016) argue that low food prices and high land and infrastructure costs are strong disincentives for these farms. However, with 'many of the "good food movement's" core ideals incorporated' in their agendas, urban farms are willing to take risks. A review of the literature documenting theoretical and empirical studies in Western Europe, concluded that peri-urban farming is strongly permeated by social objectives and that, in addition to food production, models like care farming (aimed at improving the mental health of participants through gardening) are on the rise (Zasada, 2011). In light of the considerable constraints that the urban context presents, it is not surprising that farms based on typical for-profit business models are not many. One of them is Hantz Farms in Detroit, a city that has witnessed an intense activity in urban agriculture as an engine for regeneration, in a context of industrial and economic decline and the shrinking of the population (Colasanti et al., 2012). However, the plan of the company to purchase a large site in the city was met with diffidence and a suspicion that the operation was mainly speculative (Safransky, 2014). Perhaps the only examples of companies investing considerable funds in urban agriculture commercial projects are those based on soil-less technologies. Lufa Farms in Canada and Urban Farmers in Europe are each a case in point. These projects are typically urban because they are located on rooftops, occupying relatively small although intensively cultivated areas and they are being promoted as highly sustainable, perhaps not because of a direct involvement of local groups but rather because of their short supply chains, with produce sold locally.

The reason for this book to include small scale community-led and commercial soil-less projects is connected with their shared focus on the positive social impact of urban food growing as well as the mixture of commercial and non-commercial activities on which they are based. For example, in the UK, it is very common for community projects to grow edible crops to sell, develop programmes of socio-cultural activities that can generate income and seek funding opportunities (Social Farms and Gardens, n.d.; see also Chang & Morel, 2018). In France, urban market gardens (termed microfarms) are becoming an established model to produce and sell food in cities (see for example Scheromm & Soulard, 2018; Morel et al., 2017). In an attempt to design a new typology of urban agriculture initiatives, Krikser et al. (2016) identify types that mix self-supply with economic activities, and types that mix socio-cultural with economic activities. On the one hand, it is inevitable that an agricultural practice within an urban environment with high social interaction specialises differently from that operating in a rural context. Urban farmers are influenced by the rich and intense exchange of activities and experiences occurring in cities and modify their attitudes and practices accordingly, with the several facets of urban life feeding into each other. On the other hand, whether commercial or non-commercial, any urban agriculture activity is oriented towards a more sustainable production and consumption of food. However, it is also inevitable that the expansion of activities for the commercially oriented projects may result in prioritising economic returns to sustainability; small enterprises can grow and modify their business model. The balance between financial viability, economic return and sustainability is hard to strike, although it is, this book maintains, a fundamental requirement for the development of an alternative model of urban, short food supply chains. The exploration of the motivations, challenges and opportunities of all these small scale soil-less projects in which economic, social and environmental values are so closely linked, becomes therefore an exploration of the new directions that urban agriculture can take.

The small scale of these projects means that the personal life trajectory of the farmers, as the case studies in Chap. 6 show, is deeply connected with, and influences, their practices. Their cultural background, views and purposes are intertwined with their projects and show how production, which in this modernity is a concept predominantly associated with industrial processes, can be humanised and a positive part of our daily lives. These personal trajectories can also provide clues about the changing identity of urban farmers and the changing perception of the scope of urban agriculture in the cities of the Global North. With urban soil-less farmers, this practice becomes an opportunity to understand how technology – evermore associated with the artificial, non-human dimension – can be appropriated, transformed and democratised, also because the small scale allows these projects to avoid prioritising – to an extent – economic return. The argument for a higher convergence between human life, human values and agriculture, is one promoted by agroecology, in an attempt to define agricultural systems not according to neoclassical economics but rather on their capability to connect with ecology, landscape, bioregions and communities (Francis et al., 2003). An urban agroecology approach is one that deals with food sovereignty governance of food systems (Pimbert, 2017), therefore taking a political stance. Soil-less technologies are perhaps disconnected from real nature; they attempt to modify and replicate ecology, often indoors. Nonetheless, when used in urban small scale projects, as shown in the case studies, they often embrace aims that resonate with urban agroecology, bordering on the political sphere. In addition to food sovereignty (enabling control of the food production and consumption cycles), the advantage that an intensive, space efficient food technology can generate in absolute terms is to reduce demand for agricultural land and potentially enable conversion of abandoned agricultural land (Caputo et al., 2020; Despommier, 2010). Motivations of urban farmers starting soil-less small scale projects can therefore coincide with the drivers of urban farmers committed to a system of food production that has a positive impact on local ecologies as well as global environmental amelioration. However, as the case studies will show, this aim is not always clear to soil-less urban farmers.

The investigation presented in this book focuses on three soil-less techniques: hydroponics (a technique enabling plants to grow with their roots in a nutrient solution); aquaponics (a closed-loop system in which fish are farmed in tanks and the water enriched with nutrients coming from the fish's excrement is used to feed plants growing in a hydroponic unit); and mushroom farming (in which mushrooms are grown indoors, in different media, including straws and spent coffee grounds). These are not the only soil-less techniques of cultivation; aeroponics, a technique in which roots feed from nebulised water with nutrient, is, for example, a variant of hydroponic technology. However, these variants are generally quite sophisticated and sometimes experimental, whereas the projects documented in this book rely on rather simple soil-less technologies. The investigation is mainly qualitative and based on case studies. 12 projects (together with two educational projects) were visited between 2018 and 2020, in seven European countries: the UK, Spain, Italy, Germany, Belgium, the Netherlands and Sweden. The choice of locations was the result of a long search on the internet and through personal work contacts. Some of the projects initially identified declined the request for a visit or did not respond to it; others – on closer examination – were not suitable for the scope of this book. A list of the projects that were contacted or any other project that was identified in the course of the study on which the book is based is included in Appendix A. It gives an approximate scale of the size of the small-scale soil-less phenomenon, which, as noted already, is rather small. However, it is highly likely that a large number of initiatives escaped this search. For example, initial contacts with urban agriculture experts in Italy did not help identify soil-less projects. The one included in the book was accidentally found during a visit to a trade fair in Italy.

The urban farmers leading these projects were interviewed and asked to provide data on the productivity and resource use of their projects. Not all farmers could provide sufficient or reliable data, mainly because good yield and efficiency are not parameters that matter for all. For some projects, social engagement and educational outreach are a priority; gathering quantitative evidence would take time and efforts away from their main aim. The dataset collected is therefore small. Yet, it can give a broad indication of the potential of urban soil-less small-scale projects. It must be stressed, however, that the main merit of the book is the documentation of the stories, opinions and motivations of the urban farmers, which compose a picture useful to comprehend the latest evolutions of urban agriculture and can be a resource for future studies in this area. Across the book, food technology, the evolution of urban agriculture, its political and social role, and urban nature – which have been outlined above – will be elaborated and discussed within the perspective of soil-less technology.

The following chapters offer an historical, theoretical and empirical investigation. The theoretical part starts in Chap. 2 and provides a reflection on the factors driving the transformation of urban agriculture over the last two decades, resulting in a multiplication of diverse types and models, including soil-less urban agriculture. It argues that, because of the urban context and the actors involved, urban agriculture has a propensity to experimentation, which leads to a continuous process of social innovation. Indeed, experimenting (with new social models, new organisational structure, new business models and new techniques) is one of the characteristics of this practice. The following chapter offers a reflection on technology and society, which helps locate the adoption of food technologies by urban farmers within a context of a society ever-more dependent on and, at the same time diffident towards, technology. Barriers to comprehend and appropriate technology are at the root of this diffidence, resulting in new perspectives which try to close the gap between specialist knowledge and laypeople. Concepts such as alternative technology and frugal innovation have contributed to close this gap and have been applied in some of the projects here documented. The historical part (Chap. 4) includes brief outlines of the evolution of the three soil-less technologies utilised in the case studies: hydroponics, aquaponics and mushroom farming. It also describes some of the several techniques available for each one of these technologies.

The empirical part includes the case studies with a description of the project and a summary of the data collected, when these were available. These data and the informal interviews of the farmers provide material that is analysed in a subsequent, dedicated chapter (Chap. 6). The analysis focuses on the motivations that led to the adoption of soil-less methods for each project, the actual outcomes (i.e., achievements), the perceived and effective environmental efficiency and, finally, the productivity, which may not only refer to food but also to social benefits. An investigation on the internet into communities of soil-less farmers is developed in a dedicated chapter (Chap. 7), complementing the case studies. The internet augments opportunities for groups to exchange information or to promote – or advertise – innovation. It can also be used as a tool to trace the magnitude of interest of the internet users about soil-less technologies, as well as the reasons behind such an interest (e.g., commercial and private use). The chapter identifies the search words of internet users looking for aquaponic and hydroponic systems and uses these words to understand what drives them in their search. It also uses the search words to identify 'YouTubers' (i.e., users who broadcast videos on YouTube) communicating soil-less projects via their videos and, again, motivations to start such projects. It is also useful to see that some of these projects have numerous followers, which in turn can give a clue as to the scale of interest that soil-less is attracting.

These two chapters are preceded by a methodology chapter (Chap. 5) and followed by Discussion and Conclusions (Chaps. 8 and 9). The methodology chapter provides background information justifying the selection of the case studies and explaining criteria for their evaluation. Chapter 8 provides such an evaluation and elaborates on quantitative and qualitative findings. Finally, Chap. 9 brings together findings and threads of discussion, answering questions formulated in this introduction and tracing possible directions for the future development of small-scale soil-less urban agriculture.

Chapter 2
Recent Developments in Urban Agriculture

Abstract Over the last few decades, the world of urban agriculture has considerably changed and diversified, attracting the interest of researchers, practitioners, and policy makers. The purpose of this chapter is to trace the recent fast evolution of urban agriculture in Europe and North America, which points to a propensity of urban farmers to reinterpret their approach to food production in response to changes in society. Three new trends are identified and described: 'community-based urban agriculture', which captures the shift from a practice predominantly carried out by individuals to one in which community groups manage farms and gardens to grow food and provide a service to the local community; 'metabolic urban agriculture', a term representing the attempt of farmers to root their practice within the material and social flows of cities; and 'experimental urban agriculture', representing projects that experiment with technologies and a circular economy approach to food production. These new trends coincide with a modification of the demographic profile of urban farmers which tends to be younger than before and possessing new skills, necessary to engage with volunteers and the general public, when relevant, or to deal with technology. The chapter frames the rising interest of soil-less technology within this recent dynamic transformation of urban agriculture.

The world of urban agriculture, the many ways and models in which this phenomenon manifests around the world, has considerably changed and diversified over the last few decades. A useful diagram that was included in *Second Nature Urban Agriculture*, a book edited by Viljoen and Bohn (2014), indicates 1978 as the year in which an acceleration of new groups forming and initiatives promoting urban agriculture started. In fact, in 1978, the website of Urban Farmers went live, which was probably one of the first attempts to initiate a public debate on this topic and provide resources and knowledge on a practice which perhaps, until then, had not been sufficiently researched and understood. In 1996, RUAF (Resource Centre for Urban Agriculture & Forestry) was founded and the seminal book Urban Agriculture by Smit and Nasr, printed. The diagram ends in 2010, a year in which many books on this topic were published and events organised. After 2010, interest and initiatives have further multiplied. A brief, initial search on Scopus, using the keyword Urban Agriculture, shows that between 2010 and 2019, 666 articles with urban

S. Caputo, *Small Scale Soil-less Urban Agriculture in Europe*, Urban Agriculture, https://doi.org/10.1007/978-3-030-99962-9_2

agriculture as a keyword were published in peer-reviewed journals. Similarly, two COST Actions (networking programmes for researchers, industry, third sector and administrations, funded through European funds for research) on Urban Allotment Gardens in Europe (see Bell et al., 2016) and Urban Agriculture Europe, (see Lohrberg et al., 2016) were developed, both running between 2013 and 2016. Another one focusing specifically on Aquaponics followed (The EU Aquaponics Hub), running between 2015 and 2018 (see 2019b; Goddek et al., 2019a).

The growing interest on this topic within academia and the NGO sector reflects what actually happens in the real world. In Europe and North America, the interest of civil society in growing food in cities over the last four decades has steadily increased, although it is difficult to ascertain this other than through episodic and fragmented facts and figures. For example, on its website, the UK National Allotments Society measures this growing interest in terms of waiting lists. Out of 321 local authorities that responded to a survey, 67% held waiting lists, with an average of 52 people waiting every 100 plots. 36% of all the local authorities responding stated that there were plans to increase the supply of allotment plots in response to the rising demand (NSALG, 2020). The same website reports that there are approximately 300,000 allotments across the UK, compared with more than one million that were available during and immediately after World War 2 (see also Acton, 2015). In Germany, by comparison, there is a higher number of allotments, totalling around one million plots, 77,000 of which are in Berlin alone. In Germany as in the UK, urban development threatens the survival of allotments (McGuinness, 2015), which often occupy green land with a relatively high value and in inner city locations, desirable for residential and non-residential building developments.

While reasons for the fall in the number of allotments in the UK after World War 2 can be found in a reduction in demand (corresponding to a change in lifestyles, which in the second half of the twentieth century and in cities was becoming evermore detached from the rural context and culture), the reasons for the current renewed interest and increase in demand are yet to be precisely identified. There is a narrative that transpires from grey literature and websites which references climate change driving a rising demand for local food (e.g., Perry, 2017). Another reason that is mentioned in this literature is the need for citizens to take back control of a food system that has produced massive imbalances in terms of food quality, locality and fairness for farmers (see Greensgrow, 2020). In reality, reasons for an increase in interest in urban agriculture are far more complex; beyond broad motivations driving people to urban agriculture, triggering events lead to spikes of interest. For example, the shrinking population and economy of Detroit resulting in one third of residential plots abandoned seems to have caused the occupation of many of these plots with food gardens (Colasanti et al., 2012). The latest 2007 economic crisis is another event associated with the growing interest in urban food growing (Sanyé-Mengual et al., 2018). In the current pandemic crisis, reports in the media suggest an increase of individual and community food growing initiatives in the UK (Busby, 2020) and in the US (Ngumby, 2020) to increase food security, which was perceived at threat from food chain disruptions. In the UK, an online and telephone survey of 101 farms organising vegetable box home deliveries showed that sales increased by 111% between February and April 2020 (Wheeler et al., 2020). It can be assumed that other urban farms selling through this scheme benefitted from this trend. Importantly, allotments and community gardens have greatly helped users maintain acceptable levels of physical and mental health (Niala, 2020).

The insecurity and precariousness that these events generate is counterbalanced with more concrete and permanent values, such as attachment to nature and a focus on basic but universal needs such as food and food production. In developing countries urban agriculture is often still a means for subsistence (De Bon et al., 2010), although things are changing. In a study taking its cue from the concept of models-in-circulation (i.e., models that have been successfully developed in a country or geopolitical area and replicated in another one without considering the sometimes radical differences in the socio-economic context) changing attitudes towards urban agriculture are documented (Schwab et al., 2018). The promotion of community gardens by local authorities to increase community cohesion and the availability of healthy food – a model that is considered successful in some European countries – when applied in Bogota' and Medellin, Colombia (a medium-income country) generated mixed results, with some participants associating the stigma of poverty to the food coming from these gardens.

But for the purpose of this study, it is the fast evolution of the way urban agriculture is practiced in Europe and North America that is of interest. Caputo et al. (2020) provided a description of a few models that have recently become established, which move away from the mere purpose of growing food on allotments, mainly with the purpose of producing food for personal/household consumption. Changes in this conventional model of practicing urban agriculture are relevant to the topic explored in this book, since they point to a propensity of urban farmers to reinterpret their approach to food production in response to societal challenges. In fact, the adoption of soil-less food technologies by community groups and urban entrepreneurs can be understood within the perspective of a constant search for solutions to attain urban sustainability. Some of the emergent trends are briefly outlined below.

2.1 Community-Based Urban Agriculture

As noted, in Europe and North America, urban agriculture is still mainly practiced at an individual/household level (Kitao, 2005), following a model that was established, at least in some European countries, much before World War 2, when the lack of supplies forced people to dig for food in their backyard or in parks. Over the last decades, the practice of growing food has acquired values that transcend the mere desire for gardening and the pride of growing your own food and aspire to ameliorate local social conditions. These social benefits attached to urban agriculture inspire groups in very different ways that range from radical visions of alternative lifestyles to more practical aims of supporting disadvantaged groups in society. The Transition Town movement, for example, is based on a vision of a post-peak oil, resilient society, in which social cohesion is key and locally grown food fundamental (Hopkins, 2008). This view has inspired other movements such as Incredible Edible Todmorden, a group that occupies marginal open spaces in the city of Todmorden, UK, turning them into productive green spots. This project utilises food growing as a collective and community building practice, a way to 'beautify the city' and a demonstrator that society can be based on a more humane set of values (see Incredible Edible Todmorden, 2020). Another community garden in Sydenham, London, has evolved into a group providing therapeutic and vocational horticultural sessions, aimed at those who need support for their mental well-being (Sydenham Garden, 2020). Mudlarks, a community garden in Hertford, UK, supports people with learning difficulties, offering them gardening sessions (Mudlarks Garden, 2020).

There is no census for these community projects in Europe or North America, to demonstrate the scale of the phenomenon. Nomenclature also varies. The model is named community garden in the UK, a term that can characterise allotment sites in the USA. But the number of initiatives associated with food and managed collectively can perhaps be captured by the number of members registered at the national association representing community gardens and city farms in the UK, Social Farms & Gardens, which approaches 1600 members.[1] Similarly, Capital Growth, an initiative established in 2010 with the aim of supporting and facilitating the creation of 2012 group-based new growing spaces in London by 2012, was very successful and went beyond the target.[2]

One interpretation that can help understand the range of interests underpinning community-based urban agriculture is that outlined by Ioannou et al. (2016), suggesting that such interests can be grouped under social, political and environmental motivations. Holland (2004) identified the main objective of 96 community gardens in the UK. These included education, protection of an area, community development and business opportunities, in addition to the more conventional objective of food provision. There is also a more political side to community gardens, clearly captured by the Guerrilla Gardening movement (see Reynolds, 2014) which provides the following definition of the term: 'the illicit cultivation of someone else's land'. This is a definition that expresses the reaction to the privatisation and commodification of the city by those who have no access to land. They use the power of plants to colonise unused urban land and, at least symbolically, contrast private ownership and claim self-management of the public space (Adams & Hardman, 2014). Another political reading of the phenomenon of community-led urban agriculture views it as a consequence of a post-political age, in which political institutions have devolved some decision-making ability to powerful actors. In reaction to this devolution, citizens' groups are progressively taking into their hands the right to negotiate the use of public space (Certomà & Tornaghi, 2015).

Considering the scarcity of space that can be allocated from local authorities to food gardens, it is not surprising that the quest for and occupation of unused and

[1] https://www.farmgarden.org.uk/

[2] https://www.capitalgrowth.org/what_we_do/

suitable space has become a necessary step for community groups to start their food growing activities. If on the one hand, the determination of community groups in starting their initiatives can be seen as a sign of activism and also participation in public life, on the other hand this phenomenon is perceived as negative by others who see it as an opportunity for municipalities to delegate responsibility and involvement in the management of public spaces, while apparently showing openness to meet local groups' requests (McClintock, 2014). Rosol's study in Berlin (2012) shows how the support of local authorities to start a new urban farm mainly aimed at developing educational activities for children, resulted in the transfer of municipal responsibilities for public infrastructures such as parks to community groups. Beyond the positive or negative evaluation of community groups self-managing urban space, overall, community based urban agriculture can be understood as a social movement; a collective action by civil society, struggling for a just redistribution of resources (Barthel et al., 2015).

2.2 Metabolic Urban Agriculture

Ecological awareness and the perpetuation of an ecological memory – a memory that can be passed to the next generation in a rural context - are typical characteristics that can be seen in urban farmers, working on allotments (Barthel et al., 2015). However, in the urban agriculture research and practice, nature is constantly reconceptualised as its understanding evolves. One such reconceptualisation is that of 'ecosystem services', which was introduced in the late 1970s to convey the importance of biodiversity conservation. It subsequently attracted interest in economic studies since it was seen as an opportunity to quantify the value of these services (Gómez-Baggethun et al., 2010). The Millennium Ecosystem Assessment (2003) contributed to mainstreaming this term and attracted further interest from academia and policy. The link between ecosystem services and urban agriculture (and urban nature generally) was clearly formulated in a study on ecosystem services in cities, which proposed a categorisation of urban spaces that provided such services: street trees; lawns/parks; urban forests; cultivated land; wetlands; lakes/sea and streams (Bolund & Hunhammar, 1999). To date, there is abundant literature attempting to measure the ecosystems services associated with urban agriculture, with metrics that are not only pertaining to economy (Clark & Nicholas, 2013; Langemeyer et al., 2016; Orsini et al., 2014).

This new systemic view of understanding nature, and nature in relation to cities, is not the only conceptualisation that is changing the way urban agriculture is understood. Urban metabolism is in fact an emerging concept that sees the city as an entity subject to the same natural laws of living organisms and ecology, in which the intake of resources is transformed and reutilised with no waste. Analogies between the city and organisms are not new. Through history, the city has been compared to the human body (Sennett, 1996) and Marx coined the term of social metabolism to describe the flows of natural resources that enable production and economic activities in the industrial city. Marx also formulated the idea of metabolic rift, a disruption of the agricultural world – in tune with the natural world – triggered by the expansion of, made possible by, and generating, capital accumulation. Flows of resources were directly connected to the functioning of the city in an article published in 1965 by Abel Wolman, introducing the following definition of urban metabolism: 'all the materials and commodities needed to sustain the city's inhabitants at home, at work and at play' (Restrepo & Morales-Pinzón, 2018). Since then, numerous studies on urban metabolism have been developed, looking at energy flows, waste flows and more.

Analytical methods such as material flow analysis – which was originally developed in the field of industrial ecology in order to optimise inputs and outputs of industrial cycles (see Fischer-Kowalski, 1998) – have been used to map and quantify urban resource flows in order to assess their sustainability. The urban metabolic potential of urban agriculture has been assessed in some studies not only quantitatively by identifying flows of materials (Goldstein et al., 2016) but also conceptually (McClintock, 2010). The latter is very relevant to this book. It revisits the Marxist idea of metabolic rift, identifying its current ecological (e.g., ecological disruption as a consequence of industrial methods of agricultural production), social (e.g., consequences of urbanisation and industrialised agriculture on the livelihood of small farmers worldwide) and individual (e.g., alienation from labour and alienation from nature) dimensions and proposing urban agriculture as a means to address the rift in all its manifestations.

Sometimes indirectly, the latest theories or envisioning exercises in urban agriculture reflect the idea of a metabolic urban agriculture and demonstrate a desire to step up their ambitions, when compared to earlier studies. Urban agriculture is no longer a means of meeting basic needs (whether these are limited to subsistence in some developing countries or healthy lifestyles in countries with a mature economy) but rather to transform society. The concept of Continuous Productive Urban Landscapes (Viljoen et al., 2012), for example, merges urban ecology and urban design theories, and suggests mechanisms to progressively create green corridors that are edible and run across cities. Another conceptualisation linking urban agriculture practices with urban systems is the above mentioned *Zero-Acreage Farming*, a term coined by Thomaier et al. (2015), portraying an urban food production and supply system composed of zero-mile farming types that include rooftops and indoor farming. These theoretical and analytical approaches or metaphors offering or based on visions to resolve the metabolic rift, imbue much of the thinking of community groups and – to an extent – part of civil society. Together with the circular economy and 'de-growth', these scientific theories now inform the way we perceive society and the environment. By entering in the common parlance, they also influence the motivations of activists and civil society groups undertaking any course of action. The healing of the metabolic rift through urban agriculture, as McClintock suggests, can be seen as the driver of movements such as those mentioned above (e.g., Transition Towns and Incredible Edible Tordmorden), whose vision sets food growing practices as one of the stepping-stones enabling new societal arrangements. This leads to the idea that urban agriculture can deliver benefits that are no longer partial (e.g., food for gardeners and benefits for the local biodiversity and climate) but rather absolute (e.g., urban resilience, circular metabolism of urban resources and reduced need for more agricultural land). In this view, urban agriculture becomes systemic, and the quantification of its benefits goes much beyond the place, neighbourhood or city in which it is practiced.

A brief glance at the positioning statements of some community projects, which can be seen on their websites, confirms these assumptions. Growing Communities, a community group that organises Community Supported Agriculture and manages some food gardens, states: 'we are building a better food system that's fair to farmers, kind to the planet and great for all of us' (Growing Communities, 2020). Veni Verdi, a French association that manages community gardens and micro-farms, declares that its objective is 'to create gardens in an urban environment that have an impact on our Environment, our Society and the Economy' (Veniverdi, 2020). Prinzessinengarten, a temporary garden in a central area of Berlin earmarked for development (i.e., nomadic garden) goes a step further and suggests that this community garden can become 'a miniature utopia, a place where a new style of urban living can emerge, where people can work together, relax, communicate and enjoy locally produced vegetables' (Prinzessinengarten, 2020).

Land is another resource that is part of the flows feeding urban agriculture. As inner-city land is scarce, community groups colonise abandoned industrial yards (see Edible Eastside (2020) in Birmingham) or occupy spaces on a temporary basis as noted above for Prinzessinnengarten (see also Skip Garden in London – Global Generation, 2020). Nomadic gardens present the advantage of being deployed anywhere, thus allowing any urban plot to become part of a green infrastructure that can be reconfigured because it is mobile. In this way, food production it is not exclusively linked to its traditional location (green areas), hence creating green infrastructure without necessarily expanding the surface of green areas. Given that there is a disparity in access to quality green space between communities of different socio-economic status (Rigolon, 2016; Rigolon et al., 2018), nomadic gardens can, at least for short periods of time, reduce this disparity. There are drawbacks to nomadic gardens, as efforts to construct a growing space, a sense of identity and a place-based network can be inhibited when the garden is moved or disappears. However, nomadism can be interpreted as another manifestation of a metabolic approach, which suggests constant transformation and circularity of resources.

2.3 Experimental Urban Agriculture

The opportunities for growing food in an urban context have attracted interest from many areas, including business and research. This has led to the development of experiments that can potentially generate significant innovation. Some enterprises are investing in urban agriculture by starting production at a scale that can potentially have an immediate impact on food chains. A case in point is Lufa Farms in Montreal, an enterprise with three indoor hydroponic rooftop farms of 4000, 2900 and 1300 m^2 respectively. Lufa Farms claim to produce no water waste, to use no synthetic fertiliser and pesticide, and to compost all food waste. The greenhouses are digitally monitored and grow sufficient produce to feed more than 10,000 people in the Montreal area (Halais, 2014). Food is sold through a vegetable box scheme, with customers buying online and fetching their box from drop off points in the city. Lufa Farms is not the only enterprise investing in this field. The Brooklyn Grange rooftop farm has been described as one of the largest of this type. The enterprise that manages the farm expanded and at present occupies three rooftops totalling 22,200m^2 (5.5 acres) in Brooklyn and Queens. On its website, it claims to harvest 181,500 kg per year of produce and to have hosted 50,000 young people for educational trips to date (Brooklyn Grange Farm, 2020). To make a high-tech food enterprise profitable, however, is challenging. The ambitious aquaponics farm designed by Urban Farmers on the rooftop of an empty office block in The Hague (including a 1200 m^2 hydroponic farm in a greenhouse and a floor below hosting the aquaculture component) was closed soon after its opening because it was deemed commercially unsustainable (Hortidaily, 2019).

When environmentally efficient, commercial urban agriculture can be seen as an alternative to a resource-intensive industrial agriculture and a more resilient approach to a globalised food trade that seems to be increasingly vulnerable to climate change, local conflicts and politics. Many enterprises, like Brooklyn Grange, include educational programmes in their activities, thus demonstrating their awareness of their social obligations. However, there is a risk that, in a scenario in which urban agriculture scales up production and significance, commercial and speculative logic is prioritised, as the Hantz Farms precedent, mentioned in Chap. 1, suggests. Effective urban policies that can facilitate the trialling of new models and their consolidation when they conform to principles of environmental and social sustainability, and equitable access to resources for all, are necessary.

Effective urban policies can protect a space and role for a community-led urban agriculture while allowing urban food enterprises to thrive. In this perspective, it is reassuring to see that synergies can be created between local administrations and urban food enterprises. In Oberhousen, a city with 200,000 inhabitants in the German Ruhr area, a new headquarters for the job centre was designed with a greenhouse rooftop hydroponic farm. The greenhouse is highly technological and includes an independent entrance and an educational space for visitors who can access the facility without intruding on its operations or sterilised environment. The ALTMARKTgarten occupies more than 1000m^2 and will provide local food while at the same time offering 160m^2 of its facilities for R&D activities. Investigations on the optimal use of the building's waste resources (heat and water) and correct and efficient light exposure will be carried out (Hortidaily, 2019). This innovative concept was developed in collaboration with InFarming, a German group that designed and implemented the experimental Water Roof Farm in Berlin, which utilises wastewater (rain, grey and black water) from the residential building on which it is hosted and produces fish and greens (Roof Water Farm, 2020). The close collaboration between research, policy and enterprise seems a promising direction to design and test options which, if successful, can inform urban policies and funding in particularly sustainable and economically viable sectors.

Academia is also experimenting with new forms of urban food growing to improve its sustainable performance, yield and people's acceptance. In Spain, for example Universitat Autonoma de Barcelona (UAB) is using the rooftop of 'Institut de Ciència i Tecnologia Ambientals' (School of Science and Environmental Technologies) as a food garden, fully integrated in the building's metabolism through the exchange of water, energy and CO_2 flows (Sanyé-Mengual et al., 2014). Along this line of research, AgroParisTech (Paris School of Agronomy) started on its roof T4P (Toits Parisiens Productifs), a rooftop food garden project which uses urban waste as substrates (Grard et al., 2015). The University of Greenwich, London is equipped with over 14 roof gardens and a aquaponic unit as facilities to teach aquaponics in one of the few specialised university courses on this subject.[3] These facilities and experiments are not only instrumental in identifying technologies and production techniques that are particularly suitable for urban farmers but are also likely to disseminate knowledge beyond academia, reaching associations and groups working in the urban agriculture sector.

Experimenting is not relegated to academia, R&D organisations and enterprises. Community groups are actively engaging with it too. A case in point is Prospect Farm, a non-profit project in Brooklyn which, together with growing food, is also concerned with soil-remediation, testing processes that can lower the concentration of lead in the soil in which they cultivate with several concentrations of compost in their beds (Prospect Farm, 2019). In partnership with a community garden in London, LEAP, a company producing urban micro anaerobic digesters, that is, digesters that can be installed and operated in an urban context, is experimenting with a by-product of this process. The decomposition of waste food produces biogas and a mixture of microbial biomass and undigested material (i.e., digestate), the liquid fraction of which is rich of nitrogen and potassium (Monlau et al., 2015). LEAP is experimenting with how different solutions of digestate and water can be used as an effective organic fertiliser in hydroponic systems. The experiment is not exclusively researching how to produce affordable, low carbon and powerful organic fertiliser but more importantly how urban food production based on a circular economy can be designed and implemented. Circular economy is an economic model decoupling economic activity from the consumption of finite resources and designing waste out of production systems. It is underpinned by the transition to renewable energy and materials (Hellen McArthur foundation, 2021). One of the circular economy systems implemented by LEAP includes the café that operates within the community garden, which provides waste for the digester; the digester providing gas that operates a gas boiler heating a poly-tunnel; green waste from crops harvested in the poly-tunnel; and the digestate used as fertiliser. In these experiments, LEAP is supported by researchers from different UK based universities who

[3] https://www.gre.ac.uk/research/activity/las/aquaponics

provide expert knowledge, working together with this small company and the community group that runs the garden.

Experiments in agroecology are also worth a mention, with urban farmers using organic methods of production based on a combination of modern technologies and traditional horticultural techniques (Morel & Léger, 2016). These experiments and the others mentioned in this section demonstrate the propensity of urban agriculture to experiment with new arrangements in the social, environmental, technological and economic spheres. This is also possible because of the demographic and cultural profile of urban farmers which has changed considerably from the 1950s onwards.

2.4 Profiles of Gardeners/Farmers

Data on the demographic profile of urban gardeners and farmers are scarce, fragmented and not completely reliable. There is no specific study providing such data, although some information - rather inconclusive – can be gleaned from a few documents and studies. A 1998 UK Parliamentary publication from the Select Committee on Environment, Transport and Regional Affairs on the demand for allotments reports that, while in the past the typical profile of an allotment holder was that of a retired person, things were changing. As of the year of the report, out of 250,000 allotment holders in England, 35% were over 65, 30% between 50 and 64, 29% between 35 and 49, and 6% under 35. The same report claimed that the share of women gardening on allotments had grown too, moving from 3 to 16% between 1969 and 1993. This happened in a context in which 'four fifths of adults claim to garden in one way or another and 39 % describe themselves as keen, spending as much time as possible in the garden' (Petts and World Health Organisations, 2001). There is no other formal mention of the age mix of urban gardeners and farmers in the UK that can be found in official documents and studies, but it can be assumed that the age groups and social profiles today have changed. Anecdotal evidence gathered while developing this book and before, suggests that the age of people involved in the activities of community gardens is rather low, varying from 30 to 60. A 2019 small survey conducted in 9 community gardens in London with 46 respondents, including volunteers and garden managers, shows that 10 were equal or below 30 years, 26 between 30 and 60, and 10 above 60. The same survey was repeated in 6 social farms in the USA; the 38 respondents were all below 30. This is because these American social farms are targeting this particular age group by offering training skills to young unemployed. Yet, the fact that such offers are available to young generations shows that an interest within this group has been identified.[4]

[4] The surveys were conducted within the FEW-meter project, a research project funded by Urban Europe and Belmont Forum (https://jpi-urbaneurope.eu/project/few-meter/).

Unlike the UK, information is available in the US that gives a more reliable – albeit far from complete - picture of the urban farmer's demographics. A survey conducted by Oberholtzer et al. (2016) on 315 respondents who identified their farm as urban or peri-urban, shows that 'urban farmers are generally younger than the overall farming population, with the average age of 44, ranging from 21 years of age to 78'. The world of European community gardens is likely to be very different from the farms surveyed in the US, which will probably focus on production and commercialisation of the produce rather than food production and social support. As far as the former is concerned, the opportunity for a permanent or semi-permanent job or business opportunity is likely to drive choices. In line with this assumption, an article on the Washington Post (Dewey, 2017), suggests that there is a 'growing movement of highly educated, ex-urban, first-time farmers who are capitalizing on booming consumer demand for local and sustainable foods'. If the article correctly reports job trends, it can be assumed that a share of this generation of highly educated jobseekers will start – or has started – urban farms. A survey on a small sample of urban farmers (n = 22) in Paris, reached a similar conclusion; in the sample, all farmers had university and doctoral degrees, although limited prior knowledge in agriculture or horticulture (Aubry & Daniel, 2017).

The situation is different in Japan, where urban agriculture is deeply embedded in the urban fabric and has been practiced for some time. Similar to the ageing of farmers in rural areas, urban farmers are ageing too, posing a risk to the survival of agriculture and national food security (Moreno-Peñaranda, 2011). Conversely, a survey including 250 respondents to questionnaires distributed in two cities in Zimbabwe (i.e., Bulawayo and Gweru), in which urban agriculture is mainly a subsistence and commercial activity, shows that no predominant age group can be identified, although most farmers were found in the 21–40 and 61–80 age groups, mostly consisting of women (Hungwe, 2006). This is hardly surprising, since, in 2020, the largest share of the population in Africa (67.4%) was less than 30 years old (Statista, 2020a).

The challenge of an ageing population of farmers is a shared concern amongst European countries. The EUROSTAT farm structure survey (2015) points out that across the EU, a majority (57%) of family farms are managed by farmers above 55 years. This data is not disaggregated between urban, peri-urban and rural farms, but an analysis of the 2010 data from Lombardy, Italy, shows a larger share of farmers below 40 in urbanised areas, perhaps indicating a trend of generational renewal in urban farming. The report summarising these findings, commissioned by the Policy Department for Structural and Cohesion Policies of the European Parliament (Piorr et al., 2018), also mentions that digital technologies are increasingly important in agriculture and urban agriculture. Vegetable box schemes and food coops, for example, are enabled by information technologies, marketing and coordination. With an expanding role for digital technology, it is possible to predict that, depending on the particular model of urban agriculture, the utilisation of advanced technologies such as robotics, sensors and remote control will soon be trialled. With younger generations becoming increasingly attracted by gardening, openness to testing new technologies, familiarity with digital tools and even scripting is gathering pace. Commercial enterprises in urban food production rely on information technologies not only to control environmental conditions of, for example, indoor farming, but also to reach customers and deliver produce. This requires a different knowledge, set of skills and, above all, confidence in the idea that technologies can be used by all, rather than by experts only. This happens not only because new generations are better informed about technologies but also because of the progressively increasing involvement of researchers and community groups in co-developing research. LEAP, for example, relies on the collaboration of PhD students who, either on a voluntary basis or within funded projects, collaborate with them and in the process transfer knowledge. It is a process of 'devolution' and 'dissemination' of knowledge that enables a more democratic use of science.

If urban agriculture can hold connotations of anti-modernity to urbanites who are not familiar with it, this chapter shows its dynamism and openness to pursue new directions and means to attain its aims. There is willingness to combine food growing with other activities that produce social benefits and at the same time explore new techniques, technologies and conceptual frameworks. Aims are changing too, broadening the scope of intervention from local to city-wide - and beyond - benefits. Against this backdrop, soil-less technologies are trialled. Some of the projects documented in the case studies (Chap. 6) particularly highlight the role of technology in their agenda. A case in point is the aquaponic urban farm Bristol Fish Project, which sets as its objective 'the accessibility of hi-tech urban food growing' to local communities and the application of circular economy principles (Bristol Fish Project, n.d.). Other aquaponic micro enterprises such as GrowUp in London (GrowUp, n.d.) have a similar approach in that they organise their high-tech food business with a clear sustainability and social sustainability drive (e.g., electric vehicles to deliver produce and recruitment of employees from a local charity assisting young unemployed people). Soil-less technologies are particularly suitable for the urban environment because of their space and resource efficiency. Yet their utilisation poses new issues and challenges such as organic certification which is not extended to soil-less produce, new skills and knowledge necessary to construct and manage soil-less units and its 'artificial' characteristics which may discourage those who practice urban agriculture in order to be close to nature. Yet, the changing profile of urban farmers suggests that they are open to technological innovation. However, as a whole, society has an ambiguous relationship with technology and science. The following chapter discusses this point to finally introduce approaches to technology which explain why and how urban farmers may accept soil-less technologies, and appropriate and reinvent such technologies in their projects.

Chapter 3
The Broader Debate on Science, Technology, Society and Food

Abstract To understand the uptake of soil-less technologies within small scale urban agriculture projects and enterprises, a broader reflection on the current state of technology, society and food is necessary. The chapter traces the evolution of the relationship between people and science over the last decades. Soil-less technologies can be resource and space efficient but their contribution to urban ecology and their underpinning of values of social support and community building, which typically motivate urban farmers is unclear. This chapter helps demystify such issues. Firstly, policies and attitudes about the promotion of science and technology to the general public in Europe and North America are briefly outlined. Such policies have gone through several phases, moving from initial attempts to close a scientific knowledge gap in society, in the belief that this would have led to mitigate diffidence towards governmental investments in technology, to the recognition that science had to reapproach society in order to be in line with people's needs. Secondly, concepts such as alternative technology and frugal innovation are explored, which were generated in response to such a recognition. Finally, these concepts are transferred to the field of food production technologies, proposing the latest adoption of hydroponics, aquaponics and mushroom farming within small scale urban agriculture projects as examples of frugal innovation.

3.1 Overview

In a book that investigates the use of soil-less food technologies in urban agriculture projects, a broader reflection on the current state of technology, society and food is necessary in order to correctly frame this phenomenon. To this end, the chapter traces the evolution of the relationship between people and science over the last decades. Topics presented here go beyond the specific focus of this book to then apply some of the latest understanding of science in society to that of small scale soil-less urban agriculture. Broadening the view on science, technology, society and food allows these urban agriculture projects to be put into historical context.

Research into soil-less technologies has a long history (see Chap. 4) but only recently has this research focused on and supported the industrial development of

S. Caputo, *Small Scale Soil-less Urban Agriculture in Europe*, Urban Agriculture, https://doi.org/10.1007/978-3-030-99962-9_3

equipment and systems for soil-less cultivation, enabling the design and construction of farms at all scales, and for purposes that range from self-supply to large scale production. The potential to scale up soil-less production has consequences in terms of quantities of food produced and made available on the market, impact on diets due to out of season crop availability, new business opportunities and more. But what are the consequences for an understanding of urban soil-less food production and the perception of this type of production from those community groups and small enterprises that embrace it? How does this technology interact with the values, objectives and perceptions of urban farmers who are typically motivated by urban sustainability principles and efforts to enhance nature in cities? An understanding of the history of the science and theories that analyse the feedback loops between technology and society can help answer these questions.

A common philosophical view that permeates Western civilisation and that was fully formulated during the Enlightenment is that the study of nature allows the overarching laws that govern all phenomena to be identified. In turn, laws can be harnessed and utilised for the benefit of society (Corner et al., 2013). A vision of the world as governed by an identifiable set of laws implies a deterministic, cause-and-effect approach to understanding nature, which, arguably, has generated great advancements in science and technology. However, the paradox of a science based on rational, universally demonstrable facts and therefore above individual judgment is that such a vision requires shared interpretations of facts, which are in practice hard to obtain. The imposition of the Western view or rational thinking on other cultures is one of the consequences of such a deterministic vision and the belief that scientific knowledge is objective (Russell, 2015).

Such a view of the world and the way we develop an understanding of things has generated other consequences. Scientific mentality and positivism – in the interpretation of the Frankfurt School, today shared by many scholars - has permeated the realm of policy and decision-making too, resulting in the assumption that a rational approach rather than, for example, shared values, must drive policy (Collin & Pedersen, 2015). Inevitably, such a technocratic way of governing has ‘left little place for democratic control of technology’ (Feenberg, 2017). Marcuse introduced the idea that liberalism relies on science and technology and utilises them as an opportunity to pursue rational choices rather than those that would be taken if a social agenda were to drive economic and political decision-making (Feenberg, 2017). Participation of the common people in this technology driven regime is negated. However, determinism in science is not the prevailing paradigm today; the acceptance that natural phenomena are much more complex than those that physical laws can explain, and that physical laws themselves are often formulated by reproducing and observing phenomena in artificial – rather than real - conditions such as laboratories has opened up the way to theories based on complexity and on systems thinking, which attempt to reproduce how nature operates, with multiple factors acting concomitantly (Capra, 1996). The shift towards an ecological – rather than mechanistic – view of the world has also had an impact in social sciences. A case in point is Herman Daly (1996), who proposes to consider the limits to the availability of resources within economic theories, which typically ignore such limits when explaining the behaviour of markets or the determination of the value of goods. Earlier on, Herbert Simon demonstrated the need for economics to learn from psychology studies in order to better understand decision-making processes of agents operating in a competitive market, which may not be as rational as classic economy theories claim (Simon, 1966).

The definition of and relationship between science and technology is problematic too; indeed, thinking of the latter as applied science can be misleading (Sismondo, 2004). Brey (2018) defines technology as the 'products of engineering design: devices, systems, procedures and methods that are developed by engineers and used in society for practical ends'. But not all products generated through engineered design are the result of scientific research; some can be generated through lines of knowledge tradition, for example through craftsmanship, not informed by science or a structured process of development. Technology can be also understood as technical knowledge (Bergek et al., 2008). With regard to soil-less technologies, as the case studies will show, scientific research helped develop them, but urban farmers can transform these technologies and use them for several purposes, sometimes developing innovative solutions or new knowledge, applying scientific research methods to measure their efficiency.

Technology is not neutral. Highly specialised devices have been designed with a specific use in mind, therefore having an impact on particular social relationships and habits. A microwave oven is a tool that allows a very fast cooking process. It generates advantages while - at the same time - modifying attitudes in preparing meals and food culture generally. By changing cooking and eating habits, conditions are created that are favourable to the faster uptake of the tool – and the technology necessary to produce it. Once rooted, the tool can open opportunities that generate further change; the industry of ready-made meals expands, and diets adjust accordingly. Consequently, the 'microwave' tool has created a formidable barrier to the perpetuation of traditional culinary knowledge as well as the tradition of healthy meals prepared at home. Yet, while the impact of new technologies on behaviours and social norms can be traced, its evaluation is rather difficult; there are advantages to reducing the time to prepare food, including the increased availability of free time that can be dedicated to work and play and a number of new jobs created versus the disadvantages of unhealthy diets and the loss of a culinary memory. Technology therefore tends to open some possibilities while negating others (MacKenzie & Wajcman, 1999). Brey (2018) maintains that technology has a positive impact on society only when it enables or facilitates the realisation of shared values; that is, the fundamental values that are commonly perceived as desirable and necessary for the attainment of a good society (e.g., people's wellbeing). If there is a general agreement that one of these values is sustainability, then, for example, photovoltaic panels can be collectively understood as a useful technology for the attainment of such a value. This is an important proposition, since it links science and technology directly to the attainment and consolidation of shared values rather than – as the view of the Frankfurt school posits - the justification of contentious choices (e.g., nuclear energy plants rather than a distributed network of PV panels).

Technology can be seen as serving liberalism in determining the shape and scale of the market where technology-enabled goods are exchanged. With technology multiplying the opportunities to increase the size and range of what is exchanged on the market, customers have the opportunity to acquire goods, sometimes leading to enhance their agency. Feenberg (2017) notes that, through these goods, networks are formed: networks of drivers demanding better roads; networks of pedestrians demanding cleaner air; and networks of cyclists demanding higher safety. The formation of these networks is also an opportunity to exert pressure on the market and/or policy makers and, in a way, influence the evolution of technology. In this interpretation of the social-technological interaction however, the big players in the market offer new possibilities through an object and social networks form around these possibilities and their consequences (i.e., the car enabling fast movement using fossil fuel and a progressive pollution of air). Networks are reactive to the market – rather than proactive - and can only lobby for transformations of the object.

Opposing a deterministic view on the way technology is produced, Brey (2018) sees the technology-society relationship as more fluid. This is because individuals and communities change their behaviour and social norms to adapt to new technologies but the way this happens is strongly influenced by contextual factors. Generally, the utilisation and the impact of smart phones in rural areas of Africa differs radically from that we can observe in developed countries, the former being used mainly for practical purposes (Aker & Mbiti, 2010) and the former often being used for leisure. To an extent, this holds true for soil-less techniques too, with simplified hydroponics and aquaponics recommended in developing countries for subsistence (Somerville et al., 2014; Fecondini et al., 2010) and more technologically sophisticated systems utilised in developed countries as an efficient way to maximise production while reducing land and water usage. Lastly, the Actor Network Theory offers a perspective in which humans and non-humans are actors within relational networks, each with interests to accommodate (Sismondo, 2004). Power relations are at play within these networks and each actor strives to succeed. A case in point, mentioned by Sismondo (2004), is the attempt by EDF, the French energy corporation, to design a scenario that involves a widespread use of electric cars, which would set the basis for their new product to be commercialised. This interplay between technological tools, a vision of society, scientists and engineers constitutes the network in which actors operate.

Although technology pervades all aspects of human society and people are keen utilisers of technology, common perception varies and can veer towards scepticism. Such a stance has deep roots in the industrial revolution (e.g., Luddites), it manifests throughout the late nineteenth century, and persists during and after the World Wars (Irwin & Wynne, 1996). The latter were times in which there was a common awareness that science had been used to advance knowledge on weapons of mass destruction such as nuclear bombs. A UK House of Commons Select Committee on Agriculture stated that 'Scientists do not automatically command public trust' (Irwin & Wynne, 1996). Still today, science can propose solutions to problems which are not in line with the public feeling. The advice from the European Commission (2020a, b) to fell all olive trees infected with Xylella

Fastidiosa, for example, triggered great controversy and opposition amongst farmers in Italy and elsewhere, who were contesting scientific advice on the most effective approach to limit the spread of the bacterium, as this was not aligned with their approach to problem solving (Colella et al., 2019). Science proposes new directions that can change the way society understands itself; it introduces technologies that create new needs and social relationships that are not neutral and therefore can be contentious (Irwin & Wynne, 1996). The debate on cloning and GMOs, for example, poses fundamental ethical questions and a common agreement on this issue has yet to be found.

Theories provide valid support in understanding dynamic interplays and evolutionary processes within current society in terms of science and technology. But just because technology has such a significant role within society, policy making needs to deliberate on large investments in research and technology which require public consent. Hence the preoccupation from central authority to better inform people about scientific and technological progress. Attempts to bring science and technology closer to people have often relied on the popularisation and diffusion of science rather than its democratisation; science as understood by all but not produced or determined collectively (Russell, 2015).

3.2 Policy and the Promotion of Science and Technology

Studies on the Public Understanding of Science were developed from the 1950s onward, in order to identify the perception of science and scientific knowledge that people held. These studies were based on large surveys in the US and Europe, which were meant to ascertain the general level of scientific knowledge. They influenced top-down approaches to the development of relevant strategies and policies. Bauer et al. (2007) identify three phases corresponding to different directions of research into the public understanding of science. The first one, between the 1960s and 1980s, is based on surveys suggesting that a widespread literacy deficit was at the basis of a poor understanding of science and a diffidence of laypeople towards scientists. Within this perspective the promotion and attainment of higher levels of scientific literacy (e.g., encouraging students to study scientific subjects) can overcome public diffidence and increase participation. The main criticism to this stance is a paternalistic attitude of scientists and institutions, assuming that there is a widespread scientific knowledge *deficit* and deducing that it is the responsibility of people to act rather than of institutions to take a different approach to scientific research. The second phase (between the mid-1980s and 1990s) moved from an idea of knowledge *deficit* to one of *attraction*. This phase promoted a strategy aimed at convincing people that science is attractive (*the more you know, the more you love it*). Regardless of the shift, this new attitude still assumed that people are ignorant, thus feeding a negative reaction from the general public which did not help reduce common mistrust.

The last wave started in the mid-1990s and is termed *science and society*. Eurobarometer surveys show that between 1992 and 2001, interest in science declined but scientific knowledge increased, thus suggesting that public mistrust was not only the result of a deficit in scientific knowledge, since such a mistrust was manifested by people who had good levels of knowledge. Within the vision of *science and society*, the responsibility of starting a process of rapprochement between the two cannot be left to people; the scientific world must change its attitude and find ways to actively involve society in the development of science. Participation is one of the most valid methods to attain this goal. Public ongoing debates such as those on GMOs are an example of this new attitude to move towards a democratic approach to science and technology. It is important to stress that these phases influenced the top-down framing of this issue and, in turn, the identification of research agendas and connected investments. For example, priority areas within the several EU-funded schemes shifted from *Science and society,* within the EU Framework Programme 6, to *Science in society* under the subsequent funding programme (2007–13) (Adamsone-Fiskovica, 2015).

Research on public understanding of science has not only informed the broader European research agenda but also the attitude of policymaking. The report *Taking the European Knowledge Society Seriously* (Felt et al., 2007), acknowledges that there is a general unease in society towards science and its products (i.e., technology) and that this can be addressed through the governance of science (Tlili & Dawson, 2010). The realisation of this mistrust is a consequence of surveys suggesting that causes of this mistrust may lie in a shared perception that science and scientists are deeply intertwined with economic power (Prange-Gstöhl, 2016), although other surveys suggest that scientists are trusted more than scientific information itself (Yarborough, 2014). For the European Commission, the way to address this ambiguous relationship with science lies in the way science is sometimes used to take normative decisions which carry value-related implications (Felt et al., 2007).

The institutional attitude of the EU in a connected world is that information – scientific knowledge too – can be easily gathered on the web. Therefore, using it and processing it is more important than knowing (Digital Europe, 2010). In this latest EU stance, the relationship between science, scientists and people is ambivalent. With reference to IT and the web, the report maintains that the acquisition of lifelong skills enabling the use of scientific information must be facilitated by the scientific community and institutions (schools but also the media or any other organisation providing information) but people must rise to 'the challenge of a knowledge-guided society'. In other words, the theory of knowledge deficit is merged with that of participation in decision-making based on scientific evidence.

In order to address all this, the concept of Responsible Research and Innovation (RRI) was introduced as part of a new narrative underpinning the latest research programmes funded by the European Commission (Owen et al., 2012). RRI strengthens the idea of Science for Society and promotes the aim of opening up 'new areas of public value for science and innovation' as well as involving stakeholders in and making them responsible for a co-development process. Within this vision, universities, organisations and enterprises must involve the general public in the development of science meeting societal needs (Owen et al., 2012). Collaboration, from the scoping of research agendas to disseminating and implementing research findings, is key to a democratic process of development and utilisation of science. This is recognised in the Work Programme of Horizon 2020, the EU programme funding Research & Innovation projects (EC, 2020b). Living labs and co-design workshops are now part of the methods that are explicitly mentioned in research calls and that are identified as those that can lead to inclusive decisions. Many calls for funded projects require the involvement of SMEs (as well as local civil society stakeholders), which is an effective way of ensuring that scientific knowledge and the latest knowledge on technologies is also disseminated to those enterprises that are too small to fund R&D activities or invest in research generally. Although too early to know to what extent this approach is generally successful, this book documents a higher uptake of new technologies by small enterprises and lay people in the area of soil-less technology. Case studies show that in some farms or gardens, urban farmers either have an academic background or establish continuative collaborations with academic researchers. In doing so, knowledge exchange and a closer correspondence between research foci and farmers' needs are stimulated. Strategies to enhance public understanding of science have therefore produced positive impact, although the strongest impact can be seen when science is appropriated by people.

3.3 New Approaches to the Production of Technology

Perhaps as a result of this change of tone within the science and technology debate, new concepts were developed over the last two decades, which represent alternatives to the production of science and technology operated by experts only. Examples are concepts such as *grassroots innovation* and *alternative technology*, the former portraying and promoting an idea of innovation in technology as one produced by non-experts such as craftsmen and inventors, and the latter referring to an approach to the production of technological innovation which connects with human values and therefore does not prioritise market exploitation. The idea of grassroots innovation is particularly appropriate in the context of developing countries and technology deployed at a small scale. Alternative Technologies was instead conceived in rejection of large scale use of technologies such as those utilised in industrial agriculture, which promised maximised production but delivered soil infertility, water scarcity and water pollution (Pattnaik & Dhal, 2015). Composting toilets and rooftop simplified hydroponic systems (Sanyé-Mengual et al., 2015) can be seen as emblematic examples of this alternative approach to technological innovation.

Another concept exploring alternatives to mainstream technological innovation is *frugal innovation*. This is a term characterising innovation that is aimed at low-income groups both in terms of cost and needs, hence emphasising affordability, good performance, sustainability, and usability (Hossain, 2017). In developing countries, frugal innovation generally focuses on the design of hard-wearing products, using local technologies and materials and creatively redirecting existing technologies for other purposes. In a review of existing literature on frugal innovation, Hossain (2017) finds that the majority of publications on this topic document case studies within the field of healthcare, IT, transportation, energy and water, whereas there is a paucity of studies – hence, applications – in the field of food and food production. A paper from Fall and de Zeeuw (n.d.), maintains that there is a low level of technology acceptance in urban agriculture. This is partly because urban agriculture cannot rely on an established, formalised community and is not represented through associations as much as rural farming, and also because innovation developed within urban farming does not attract attention from public institutions. The paper is not published in a peer–reviewed journal, it is not dated (although it is likely to have been written in the early 2000s) and refers mainly to urban agriculture in developing countries. However, some of these statements can be extended to urban agriculture in the Global North, in that urban farming cannot rely on established and recognised governance systems or professional organisations, with some notable exceptions such as the recently constituted AFAUAP (Association Francaise d'Agriculture Urbain Professionelle)[1] and the forthcoming New York City Office for Urban Agriculture,[2] advising the Mayor and the Council on issues related to urban agriculture. Generally, urban farming is valued for its ecological and social merits, whereas it is not often understood as an area in which technological innovation is utilised, here too, with notable exceptions such as those mentioned in Chap. 2, Sect. 2.2.

Another important form of participation in developing research is through co-creation, which is a research method that has become very popular. Co-creation methods for the development of innovation are based on the assumption that the user of a product or a service is the holder of knowledge that is essential to the effective development of that particular product or service. A case in point is a patient with a medical condition, who is the real expert when it comes to the identification of symptoms of illness or effectiveness of prevention strategies (Cottam & Leadbeater, 2004). Co-creation is not a form of mere consultation, but rather a systematic method to involve users in a collective design process, therefore sharing ownership of the process itself. In many countries, the public sector encourages this practice which, according to the literature in this area, has particular relevance for the health care and education sectors (Voorberg et al., 2015).

The production of grassroots innovation increases as the scientific knowledge gap in society narrows. This has varied consequences. For example, IT technological tools and their widespread uptake have certainly supported a now common and consolidated view of the world as interconnected and globalised. Complex scientific theories and technologies of unprecedented magnitude are debated in the media. Over the first half of the 2010s, The Guardian published articles on massively

[1] http://www.afaup.org/

[2] https://legistar.council.nyc.gov/LegislationDetail.aspx?ID=4085856&GUID=59099EDA-FFFC-44BA-B150-439BCF4D3A3B&Options=ID%7cText%7c&Search=urban+agriculture

ambitious geo-engineering projects to prevent a global mean temperature rise, hence showcasing scientific developments and scientific hypotheses claiming to resolve the global environmental crisis through technology only. The prospect of engineering the planet's supporting systems is likely to be highly controversial (Pidgeon et al., 2012) and it can be perceived as 'messing up with nature' (Corner et al., 2013). But the important point is that media coverage on scientific issues can be seen as an indicator that levels of public understanding of science have been increasing. This is confirmed by the rate of enrolment in tertiary education in EU countries, which moved from 10–20% in 1970 to 50–60% in 2014 (Our World in Data, 2019). The upward trend in confidence in science and scientists is also one of the findings of an Ipso Mori survey in the UK, which suggests that the share of the population believing that it is important to know about science increased from 57% in 1988 to 72% in 2014. Similarly, uneasiness with the velocity of change in society because of scientific progress fell from 57% to 34% in 25 years (Ipsos Mori, 2014; see also European Commission, 2014).

The impact of technological tools can also be seen in the way people retrieve information and learn. A study on millennials and the use of the internet for personal communication suggests that this generation believes that they can retrieve any information with just a few clicks. For them, learning does not require formal training, which is the typical approach to learning of past generations (Kim, 2018). The Wellcome Global Monitor, a 2018 global study on public attitudes to science and health (Gallup, 2019), found that seven in ten people have a positive opinion of science and the way this can help their lives However, there is a gap between this generation and the previous one in the way science is understood, with 53% of interviewees aged between 15 and 29 believing they possess 'a lot' or 'medium' knowledge about science, compared to only 34% and 40% of those aged respectively between 30 to 49, and above 50. Levels of comprehension of health-related scientific terms and facts vary depending on geo-political areas, with only 2% of North American and most European countries' interviewees declaring that science and scientists are incomprehensible compared with 32% of interviewees in Central Africa.

All this suggests that the level of readiness to accept technology and scientific knowledge has increased over the last two decades. Media and the affordability of some technologies have contributed to this trend. People seem to be more informed although this information is fragmented and as such is potentially misleading. In a pilot study on the acceptance of soil-less methods in community gardens, a small sample of volunteers and visitors were interviewed (n = 45); 60% had prior knowledge of hydroponics although they were not aware that hydroponically grown food was sold in supermarkets and believed that such a food was not 'natural' (Caputo et al., 2020), hence suggesting that the shared idea of 'natural' food needs clarification. In a connected world, scientific knowledge is available at the click of a mouse although open access to it comes with the risk of misinformation, fragmented knowledge and a misleading framing of issues (e.g., crops grown hydroponically perceived as non-natural, while perhaps crops cultivated industrially with synthetic fertilisers are considered 'natural'). Regardless of such drawbacks, contemporary society is technology dependent, which is likely to result in greater confidence within the current and next generations, not only in being knowledgeable in scientific issues but also being able to adapt to and interact with technological tools, and, in so doing, at times produce innovation. Perhaps, this can be seen as a democratisation of science and technology, although the final decision on the direction and the use of scientific progress are still in the hands of private investors and central governments. For example, the ongoing debate on the decline of pollinators, with scientific evidence for this decline pointing strongly at neonicotinoid compounds used as pesticides (Woodcock et al., 2016) and the persistence of European regulation still allowing its use, is a case in point.

Civic participation is seen as a way to increase consensus on the substantial and complex decisions that policy must take. With science often offering a basis for these decisions, approaches that can reconcile people with science – such as citizen science - are increasingly necessary. Citizen science has been experimented in many initiatives that include the collection of data on natural resources and the environment generally (Varumo et al., 2020). Gathering data, in turn, can lead to a better understanding of the scientific issues on which a particular initiative is focusing, which can lead to forming an opinion on a particular inherent matter or policy. Citizen science includes not only the collection of data from laypeople but also forms of co-creation leading to the development of new knowledge though active involvement of people and experts (Van Brussel & Huyse, 2019).

3.4 Technology and Food Production

Understandably, technology has had a pivotal role in the way agriculture has developed through history, with both benefits and disadvantages accrued. Improvements in techniques of cultivation have enabled demographic growth over the past centuries and – directly and indirectly - changed the face of the planet. It is not within the scope of this book to include a brief history of agriculture and civilisation but the correlation between improvement in agricultural production and demographic growth has been amply studied. These improvements have been attained through the transformation of wild land into cultivated areas as experienced between the eighth and ninth century in Europe (Montanari, 1996) and through technological advancements. In turn, technologies typically enabled (and continue to do so) higher food production either by increasing the extractive capacity of land or overcoming factors such as distance from more productive areas through infrastructure and better transportation systems (Higgs, 1976). The impact of technological development has not been linear but has proceeded in leaps. For example, the global production of grains reached 1 billion tonnes over 10,000 years, while doubling in only 40 years (1960–2000) (Khush, 2001). In 2016, a global production of 1.06 billion tonnes of maize, 749.5 million tonnes of wheat and 741 million tonnes of rice, just to mention some of the most calorie-intensive crops, was recorded (FAO, 2018). Techniques for genetic improvement of crops, irrigation techniques, fertilizers, pesticides and mechanisation greatly contributed to the advances of agricultural production in the twentieth century (Khush, 2001). Potentially, current levels of agricultural production could feed the projected population of more than 9 billion in 2050 but this would entail primarily a drastic reduction in meat consumption and a change to a more balanced diet together with waste reduction (Berners-Lee et al., 2018). In fact, data available through FAO suggest that worldwide, food production (agriculture and farming) today provides 2917 kcal per capita per day (FAO, 2020b), which exceeds the average nutritional need.

According to Fitzgerald (1991) 'agricultural technology refers to the process of systematically cultivating plants and animals, including the economic, mechanical, human, scientific and institutional forces that support such activity'. Indeed, the dynamics of interaction between technology and people similar to those described above, can be observed at a much larger scale, with the diffusion of industrial agriculture and the transformation of its role within societies. The progressive industrialisation of production and processing processes are matched with a steady decrease in jobs in this sector from 1960 to today, particularly in industrialised countries. For example, in this period, in France employment decreased from 3.9 to 0.85 million and in the USA, employment fell from 4.53 to 1.95 million (Our World in Data, 2020). The contribution of agriculture to the economy, which possibly reflects the perception of its strategic relevance to countries' socio-economic objectives and policies, decreased too. Value added in terms of share of GDP worldwide has declined steadily from 7.6% in 1995 to 3.4% in 2018, with great differences between developing countries such as Ethiopia (31%) and developed countries such as France (1.6%) (World Bank, 2020). As an average, for the European Union, this share is 6% (Debating Europe, 2020).

This shows that the role of the free market and large investors is as important as the effectiveness of particular technologies in shaping food production. The proliferation of over-processed food, the reduction in crop diversity and the uniformity of produce size and shape dictated by big retailers are only some of the many factors that determine how farming is carried out today. Yet Fitzgerald (1991) maintains that industrial agriculture is a unique form of economic activity, and farms cannot be compared with factories, with the latter being permeated by technology and the former utilising machinery and engineered seeds but still strongly relying on manual work comparable to craftsmanship or cottage industry. Industrial agriculture is part of a larger 'agribusiness' and embedded within a broader food industry, with food processing relying very much on technology and generating often innovative products and processes. The transformation of agriculture as a production that follows industrial logic has influenced the selection of crops, for example, by reducing the range that is used, thus allowing efficiencies in production, land-use and distribution (Gowdy & Bavaye, 2019). The rationalisation of production and the impact on agricultural practices becomes more evident if large farms are compared with small ones, which do not have the same organisational structure and drive for efficiency. A study by Ricciardi et al. (2018) estimates that while small farms (below 2 hectares) produce 28–31% of total crop production, they account for a much greater crop diversity than large scale farms.

Policy and public investments have played a pivotal role in facilitating the transition towards industrial agriculture, following, as noted above, a rational approach which privileges efficiency and fast returns against long-term sustainability. A case in point is the Common Agriculture Policy that apparently distributes 80% of the funding to just 20% of farms (Debating Europe, 2020). Yet, despite the application of industrial logic, and control and reliability in terms of quantity and quality of outputs, levels of agricultural production and farming are insecure and sometimes unhealthy. Contamination of crops and viral outbreaks in farming can happen, originating food scares (Stuart, 2008) which undermine consumers' trust. In fact, one of the main concerns of consumers is the safety of food – rather than environmental benefits - which seems to direct their choices towards organic produce (Thomas & Gunden, 2012).

Surveys of consumer behaviour suggest that food scares such as BSE in beef, E. coli 0157 in cooked meat, and genetically modified crops are perceived by customers as threats (Walley et al., 2000). This perception translates into the idea that food produced industrially is no longer natural, which in turn assumes that what is natural is good. A survey investigating the attitude of consumers towards three types of produce: conventional, sustainable and organic (with sustainable – perhaps contentiously – defining crops grown with conventional techniques applied rationally and therefore with reduced impact on the environment), shows that organic food is considered by far the healthiest of the three methods of production. Ironically, within this definition of sustainable, sustainable food is predominately associated with environmental benefits (Thomas & Gunden, 2012). At the same time, organic is associated with natural food, as opposed to food grown with synthetic inputs, as well as locally produced food, including urban food (Printezis et al., 2017). Also, the more food is processed, the more it is perceived as unnatural, with hydroponic crops also perceived as non-natural (Verhoog et al., 2003).

These surveys can be understood against the trajectory of the relationship between science and society outlined in the chapter, which has improved yet remains precarious. In fact, there seems to be an ambiguity and inconsistent attitude towards science, depending on the particular field where this is applied. Understanding the natural world has enabled its modification. Society has greatly benefitted from this modification which has been facilitated through technological progress. In the eyes of the consumer, nature is associated with a pristine environment (Verhoog et al., 2003). However, there is very little land that has not been modified by human action. Agriculture in particular has historically resulted in the modification of the environment, encouraged, in some periods of our history (e.g., Greco-Roman civilisation), by the belief that managed land and urban environments were much more civilised – and therefore worthy – than wilderness (Montanari, 1996). If this perception has now changed and natural is perceived as a value, the fact that any form of nature is constructed and often exploited (Peluso, 2012) is still not completely understood by the general public. In this perspective, the idea of natural should perhaps be redefined and its utilisation as a condition for sustainability carefully examined. Indeed, some studies based on the LCA of organic and conventional crops suggest that the impact on the environment of the former may be higher when production (i.e., kg) rather than land (i.e., ha) is the fundamental unit of the assessment (Tricase et al., 2018; Foteinis & Chatzisymeon, 2016).

There are two main considerations stemming from this overview of the past and present understanding of technology, food and nature. Firstly, the idea of nature in relationship to food systems must be redefined. The claims of 'naturalness' of organic food may be misleading. Arguably, organic practices are key to restoring the health of soil and local ecosystems, which is an imperative for society. But organic practices need scientific research and effective technologies too and, in any case, are part of an agricultural tradition of modification of the environment for crop production. In a world where nature is lost and human activities are pervasive, the dichotomy of natural/artificial is perhaps not useful when used to define positively or negatively food or specific technologies of food production, which should be considered for their capability of attaining shared values. Thus, the indicators for sustainability of agricultural practices should perhaps not only include environmental impacts but also, for example, land use / land conversion. It is not only organic methods of cultivations that make food production sustainable but also a balance of land used for agriculture and land left to wilderness to enhance biodiversity and increase carbon sinks. In this perspective, urban agriculture and indoor, soil-less techniques can help, the latter for their space-efficiency and the former for its use of land that has been already urbanised. Balancing agriculture and conservation is a concern already embedded in the EU Common Agriculture Policy, which provides subsidies compensating farmers' for reduced income related to the implementation of environmental restrictions on their farms (Navarro & Pereira, 2015). This issue, however, is highly complex, and there is an ongoing debate about the architecture of rewilding policies and strategies that can clash with attempts to reverse agricultural land abandonment trends and facilitate high yield agriculture (Merckx & Pereira, 2015; Pereira & Navarro, 2015). The contribution of soil-less cultivation to food production and the possible benefits in terms of land use are therefore part of a broader and multi-faceted debate on the future of agriculture.

Secondly, and connected to the first consideration, diffidence towards technology applied to food production may be counterproductive from the consumer's perspective. If good technology, as noted above, is that which attains shared values, then this should be carefully considered when promoting and explaining the advantages of different food technologies. A perception of natural as non-artificial may lead to lost opportunities to deploy technology for a higher sustainability of food systems. Indeed, informing people of the perils of the impact of processes leading to the loss of natural habitat and how this is driven by industrial agriculture provides a more convincing and realistic frame to support soil-less production. This is particularly the case when it is implemented in cities, where land is scarce, and soil and air can be polluted. Instead, the frame provided by the naturalness of food in relationship to inputs risks discriminating against food technologies that may yield great advantages.

A further important reflection stemming from a discussion on technology and food is related to the recent attempts to redefine technological progress as attainable by all. The concepts of alternative technology and grassroots and frugal innovation capture a trend of democratisation of science and technology that is gaining traction. Information technologies are perhaps the clearest example of a technology that can potentially be truly democratic, at least in its capability to be utilised by users proactively; that is, proposing services, products or social arrangements that are not merely an attempt to modify existing technologies offered on the free market, but rather to invent new ones. Indeed, the internet has provided a powerful weapon to form large networks and exert pressure on important issues. The accessibility to coding techniques and how these can be learned and used by many non-experts to generate computer or smart phone applications, has resulted in a proliferation of start-ups offering digital tools and services and, in doing so, generating innovation. Platforms such as YouTube and Facebook are used to showcase inventions and small projects developed by laypeople, that often modify or build on existing technological products. Such a powerful trend of grassroots innovation is the result of a long-term centrally-determined policy of promotion of science in society, in conjunction with a saturation of technological tools made available on the free market, enabling many to tinker and modify technologies in sophisticated ways, also in the field of soil-less techniques of food production. However, technologies, information technologies in particular, have a dark side. A journalistic investigation on how technology has changed our lives (Bridle, 2018) suggests that the public use of technology is opaque, enabling control by an elite in ways that are difficult to discern or divert. The Cambridge Analytica affair and the role of social media in interfering with freedom of choice and political decisions is a case in point.

Patterns of acceptance of science and technology within society and the way these can be transformed into forms of bottom-up technological innovation are documented in the case studies. These demonstrate how deeply technology has penetrated our lives and how it has transformed people's views on food production, multiplying the options available to community groups and small enterprises in this sector. Land and basic horticultural knowledge are no longer the only prerequisites to start spaces to grow food in cities. Basic engineering and biology knowledge, and affordable equipment open a range of opportunities that can meet diverse objectives: from self-supply, to education, to commercial production. In the case studies, each one of these objectives has influenced the characteristics of frugal innovation developed by farmers, who have either co-created with experts, or self-built with low budget, or engineered sophisticated soil-less units, always adapting these technologies to their objectives and means available. Most of these farmers do not view soil-less produce as non-natural, with some of them believing it of a higher standard when compared to industrial produce, and organic. Having transcended a natural/artificial dichotomy, their aim is quality, low impact and outreach (for those who use soil-less systems as an educational tool), with food technology as an enabler. In fact, some of them do not distinguish between in-soil and soil-less produce but see these as equal options to grow food in cities. The urban setting is likely to be inducive to the development of frugal innovation and a broader understanding of food quality. The availability of knowledge, mechanisms for knowledge exchange as well as space and resource constraints that cities offer can be powerful drivers. As for in-soil urban agriculture, the urban environment is also a fertile ground for new groups to develop small scale projects, therefore offering an alternative model of growing food in the soil-less sector too, despite the often-high cost of these particular food technologies. The case study in Chap. 7 focuses on information technologies (i.e., the internet), aiming to ascertain the scale of interest in soil-less technologies of the general public as well as the motivations for such an interest. Here information technologies facilitate the spread of food technologies and help document forms of frugal innovation. The case study provides a snapshot of how knowledge about food production is communicated in the age of the internet as well as the profile of users who communicate such a knowledge.

The next chapter provides historical, technical and geopolitical context to three soil-less technologies, thus completing the backdrop against which small scale soil-less projects are being implemented. The chapter not only offers an interesting overview of the evolution of these technologies but also describes their different roles depending on the socio-economic situation in which they are deployed.

Chapter 4
History, Techniques and Technologies of Soil-Less Cultivation

Abstract This chapter outlines the historical trajectories, the technologies, and the techniques of three types of soil-less cultivation: hydroponics, aquaponics, and mushroom farming. This brief overview is necessary to frame historically, conceptually and policy-wise this sector and the motivations underpinning urban agriculture projects and enterprises that use these technologies. The historical sections provide an account of the factors that led to the development of hydroponics, aquaponics and mushroom farming. This constitutes a useful resource since an indepth historical investigation on soil-less cultivation has not yet been written. Tracing the origins of soil-less growing and the development of aspects such as effective growing media is both informative and necessary to contextualise these technologies within the history of food production. The historical sections are followed by a brief technical overview of the options and system components for each technology, with a particular focus on simplified soil-less systems, which have been largely promoted by organisations such as FAO to improve food security in developing countries. These systems are particularly appropriate for implementing the low-cost, self-build units operating in some of the case studies presented in the book. Finally, productivity, environmental efficiency, relevant policies, and market context complete the picture for each of the three soil-less

This chapter outlines the historical trajectories, the technologies, and the techniques of three types of soil-less crop cultivation: hydroponics, aquaponics, and mushroom farming. This book is not a manual and is not intended to provide specialist, technical insights. Rather, this brief overview is designed to provide details necessary to better understand the case studies and situate them within a historical, philosophical and policy context. The historical sections provide an account of the factors that led to the development of these technologies, which are rooted in ceaseless efforts to integrate nature into human habitats during the initial stages of urbanisation and, subsequently, with the advent of systematic scientific investigation from the fifteenth century onwards, in the relentless attempt to understand the laws governing nature. A dedicated historical investigation on soil-less cultivation has not been written yet and the historical sections presented here are based on information gleaned from scientific and grey literature, published over a century, which only

S. Caputo, *Small Scale Soil-less Urban Agriculture in Europe*, Urban Agriculture, https://doi.org/10.1007/978-3-030-99962-9_4

touch on historical facts. These sections will highlight topics and issues that drive the development of soil-less technologies, which are still being debated. Tracing the origins of soil-less growing and the development of aspects such as effective growing media is both informative and necessary to contextualise these technologies within the history of food production. In line with the scope of the book, soil-less technologies and options that can be used at both large and small scale (the latter promoted to alleviate hunger by organisations such as FAO), are briefly presented. Finally, productivity, environmental efficiency, current dedicated policies, and market context complete the picture for each of the three soil-less technologies.

4.1 Hydroponics

4.1.1 Introduction

Although hydroponics, both as a technology and a term, is relatively recent, the cultivation of plants in a water medium has a long history, and reports documenting different precedents of this particular cultivation can be found in the literature. One of these precedents, perhaps cited inappropriately as the first hydroponic system, is the Hanging Gardens in Babylon, built along the Euphrates River not far from the modern Baghdad in the fifth century BC. In reality, the Hanging Gardens are a series of terraced roofs where plants were grown in soil (Hershey, 1994). They acted as a precursor to green roofs, rather than hydroponics, together with other examples of green roofs that can be found in vernacular architecture, which were designed to improve the thermal efficiency of dwellings in cold climates (Jim, 2017). In these examples, roofs were cladded with sods, upon which grass and small plants grew. Instead, the Hanging Gardens were built to grow bigger plants in deep layers of soil, replicating natural conditions in a man-made environment. Another precedent mentioned in the literature are the floating islands (Chinampas) used in South America to circumvent the lack of suitable fertile soil in some regions. These islands were formed with organic matter, imbued with the water on which they were floating, offering an ideal medium for crops and using a growing technique that is vaguely comparable to that utilised for hydroponics (Morehart, 2016). Further details on Chinampas are given in the aquaponics section, given that the nutrient that plants use for growth was coming from fish. Another example cited as a precedent for hydroponics is the cultivation of cucumbers in a proto greenhouse on wheels, a device used in the Rome of Emperor Tiberius. Paris et al. (2008) report a passage from Pliny in 77 BC. The passage describes growing beds on wheels that, on wintry days, were moved under frames glazed with transparent stone, known as mica. These frames and beds belonged to the Roman Emperor Tiberius and were used to supply him with cucumbers all year round. Like the Hanging Gardens, this example too uses soil as a growing medium but introduces the innovation of mobility (beds on wheels) and controlled indoor environment (i.e., greenhouse) which are highly relevant to hydroponics.

These examples represent attempts to reproduce nature in artificial conditions through innovative solutions, which will be cyclically revisited and advanced over the subsequent centuries, driven by need or scientific enquiry. In particular, the greenhouse is perhaps the invention without which hydroponics could not exist as we know it today. An indoor environment maximising solar radiations enables ideal conditions for plant growth regardless of the season. As noted, greenhouses were known and exploited in Imperial Rome, but traces of glazed permanent structures used specifically for cultivation can be found dating back only to 1438. In the same year, a record can be found of the existence of temporary structures in Korea, protecting about 3-metre-tall tangerine trees in the cold months. These structures utilised semi-transparent handmade oiled paper windows letting natural light in. An underfloor heating system (a flue connected to a furnace) supplemented solar gains. The construction was dismantled every spring and reused every year (Yoon & Woudstra, 2007). The first modern greenhouses were built in Italy in the sixteenth century to display tropical plants in botanical gardens, with an extensive use of glass. Soon, the use of glazed greenhouses spread to England and the Netherlands, where indoor heating methods were experimented with (New World Encyclopaedia, 2019). Other sources credit Charles Lucien Bonaparte, a French botanist, as the first person to construct the modern greenhouse for growing medicinal tropical plants from metal and glass in the nineteenth century. As the engineering of this model of greenhouses reached an advanced stage, the functionality and scale of construction quickly became more ambitious. For example, the Palm House at Kew Gardens in London was constructed in 1844 and was 110 m long (Kew Gardens, 2019). London's Crystal Palace was built in 1851 for the Universal Exhibition using modular elements, and was covering 92,000 m^2. Other devices that allow indoor cultivation such as artificial lighting and indoor environmental controls, which are described in the following section, enabled cultivation in the absence of natural light through the use of artificial light.

4.1.2 Modern Hydroponics

The history of modern hydroponics is strictly linked to scientific investigations in plant physiology as well as to discoveries about plant nutrients and the cycles through which such nutrients are made available in the soil, air, and light. The precedents mentioned above highlight recurrent attempts to replicate nature out of its context. The integration of plants with buildings and the invention of glass sheds to trap solar heat are inventions that exploit natural elements and phenomena in urbanised environments. However, only an in-depth understanding of plant life cycles and their biology enabled significant scientific discoveries that led to the development of hydroponics.

At the beginning of the seventeenth century, the Belgian scientist, Jan van Helmont, maintained that water provided plants with the necessary nutrients for growth. He reached this conclusion by monitoring the growth of a willow shoot in a glass tuber over 5 years, which initially weighed 5 lb. and was planted in soil weighing 200 lb. Van Helmont used only rainwater for irrigation; upon completing the experiment, the willow weighed 160 lb. while the soil lost less than 2 oz. While he concluded that water provided nutrients, he was not capable of understanding the complete cycle, ignoring, for example, the role of carbon dioxide and oxygen (Resh, 2013). English physician John Woodward is usually remembered as the first person to grow plants in water culture in 1699 (Jensen, 1997a, b), possibly because he documented his experiments by publishing a scientific article describing his attempts to test Helmont's theory that plant matter is formed entirely from water. Woodward rejected this theory by growing spearmint plants in water only, using different types of water, including spring water, rainwater, Thames River water, and Hyde Park conduit water. Spearmint plants grew better in water with substances dissolved in it rather than in clear water; therefore, he concluded that such substances were fundamental to the growth of plants (Hershey, 1994).

Plant physiology was further investigated by De Saussure, in 1804, and by Boussingault, in 1851. De Saussurre presented the hypothesis that plants were composed of chemical elements obtained from water, soil, and air. He determined the amount of carbon absorbed by a plant from air and ascertained that the plant increased its mass proportionally to the amount of carbon fixed (Hart, 1930). This was confirmed by Boussingault, who grew plants in several media, including sand, quartz, and charcoal, with the addition of chemical substances. His experiments, he maintained, proved that water was functional to providing hydrogen which, together with carbon and oxygen supplied through air, formed biomass. Nitrogen and other mineral elements were also necessary to plants' functioning (Resh, 2013). In fact, while De Saussure contributed to identifying the carbon dioxide cycle, Boussingault did the same with nitrogen by discovering the presence of nitrogen in all plants. The ameliorative action of legumes for soil fertility has been known since ancient times. His discovery about nitrogen content and its importance in plant growth helped explain the role of legumes, as well as the way fertilisers affect nitrogen accumulation in the soil and have an impact on its fertility (Aulie, 1970).

The identification of these plants' nutrients and their cycles paved the way for the discovery of other specific elements which are necessary for plant growth. In 1860, the German scientists Sachs and Knop identified six elements which plants need in relatively large amounts, called macronutrients. They did so by experimenting with plant roots immersed in a water solution containing salts of nitrogen, phosphorus, sulphur, potassium, calcium, and magnesium (Resh, 2013). They also discovered that these elements must be complemented with others, absorbed in minor quantity by plants (i.e., micronutrients). With macro and micronutrients identified, Sachs and Knop were able to develop a formula for the soil-less growth of plants, which could be used to enrich water with the correct combination and proportion of nutrients. Over the following years, other scientists such as Tottingham (in 1914) and Livingston (in 1942) developed variations of this formula, testing them on plants and comparing their properties and efficacy (Arnon & Hoagland, 1944). The method devised by Sachs and Knop was termed *Water Culture*. They used sand sawdust or a cloth as a medium for the seed to germinate. Seedlings were subsequently transferred to the nutrient solution enriched according to the formula. A shallow glass container was filled with the nutrient solution in order to avoid soil absorption by porous materials (Arnon & Hoagland, 1944).

Building on the knowledge accumulated on plant physiology, major advances in soil-less technologies happened after the 1920s. In those years, many experiments were carried out on plants growing in jars filled with water or sand (Stuart, 1948). However, a real change in the history of soil-less cultivation occurred around 1925 in the USA, when experiments were initiated to find alternative solutions to cultivation techniques in greenhouses, which had the drawback of requiring abundant and frequent inputs of commercial fertiliser (Resh, 2013). Researchers in agricultural experimental stations investigated the possibility of using an artificial soil made of inert aggregates imbued with nutrient solution (Jensen, 1997a, b). In the 1930s, the New Jersey Agricultural Experiment Station and California Agricultural Experiment Station used sand as a medium and developed a sand culture method (Jensen, 1997a, b). Sand was used as an inert medium due to its mechanical properties similar to soil and the possibility of improving aeration through an appropriate dimension and mixture of particles. This approach presented some drawbacks such as accessing and monitoring root development during growth and the difficulty of thorough removal of root debris and other impurities. Nevertheless, sand was for long considered a good media although other more potentially inexpensive options were tested such as gravel, cinders, and burnt clay (Arnon & Hoagland, 1944).

But the merit of developing a real system for the purpose of industrial food production is generally attributed to an American plant physiologist at the University of California, Berkeley, William Frederick Gericke, who, in 1929, developed a more effective mixture of elements that could provide sufficient nutrition for plants when dissolved in water. He initially called this system *Aquaculture* and subsequently named it *Hydroponics* when he realised that the former referred to fish farming techniques (Hershey, 1994). Hydroponics was derived from the Greek word for agriculture, *Geoponics* (*geo* for earth and *ponos* for work). Hydroponics (*Hydro* for water and *ponos* for work or labour), reasoned Gericke, was a good substitute to aquaculture since, given its analogy with geoponics, it possessed an economic connotation which chimed well with the attempts to utilise the technique for industrial production (Hindle, 2012; Gericke, 1940) rather than research only. Over the following years, the news that a system had been devised to grow food in the absence of soil captured the imagination of enterprises and the general public, resulting in many people requesting further information from the University of California about the technology developed by Gericke. Gericke, however, was reluctant to divulge his findings on the grounds that these were the result of experiments conducted privately, not at the university, and that the system was not ready to be made public.

To resolve this dispute, the University of California appointed two scientists with the task of studying and evaluating Gericke's discovery. Their conclusions diminished the importance of Gericke's discovery and his claims of having developed an alternative system to in-soil agriculture, which they found overly optimistic (Hershey, 1994). Following this assessment, Gericke left the University of California, but his work is documented extensively through articles and books. In his manual, published in 1940, Gericke clearly stated the main aim for hydroponics in the following paragraph: 'Hydroponics is based on the theory that all the factors of plant growth naturally supplied by the soil can be coordinated artificially by the use of water and chemicals into a crop-production method capable of competing with agriculture' (Gericke, 1940:1). In the introduction to the book, he emphasised the great interest that his research raised, although he acknowledged that advancements were needed for this technology to become competitive with industrial agriculture. Yet, he showed great faith that hydroponics would offer a solution to higher demand for agricultural land, especially in countries with limited land availability and low soil fertility. As a follow up to the dispute between Gericke and the University of California, in 1950, a publication on hydroponics was authored by Hoagland and Arnon (1950), the scientists who assessed Gericke's work at the University of California, which contained detailed findings on their studies. This publication was well received by the scientific community.

In those same years, the magazines 'Business Week' and 'Time' reported that a hydroponic unit was under construction on a small island in the Pacific Ocean, Wake Island, a stop for Pan-American Airways flights to refuel. In that unit, in tanks of nutrient solutions, beans and tomatoes were grown that fed the airline's staff and crew (Hershey, 1994; How Stuff Works, 2019). This hydroponic production was part of an attempt by the US Army to test self-sufficiency in extreme conditions such as infertile land, scarcity of water, and scarcity of other resources. Hydroponic units of different sizes were built in places presenting these conditions such as Ascension Island, in the Atlantic; Atkinson Field, in British Guiana; and Iwo Jima, a Japanese volcanic island, with some of these using desalinated water. The hydroponic unit in Ascension Island was reported to grow 94,000 pounds (42,600 kg) of lettuce, cucumbers, tomatoes, peppers and radishes. The one in Atkinson Field did better and produced 234,337 pounds (106,500 kg). Two other hydroponic plants covering a total of 80 acres were planned in Japan for the US troops in a remote location which could not be easily accessed by transport, and where the soil was believed to be highly contaminated. The plant was in the open air, with one unit only located in a glasshouse (Stuart, 1948).

The recent history of hydroponics has seen substantial advancements also in the growing media. Within a perspective of optimisation of the hydroponic system for industrial use, testing media in terms of effectiveness and affordability became very important. In 1928, Robbins at the New Jersey Agricultural Experiment Station began testing sand as a suitable growing medium (Stuart, 1948). At the same time, at the New Jersey Station, Biekart and Connors experimented with carnations and found that the use of sand and nutrient solution could result in a flower production at least as intensive as in well fertilised soil (ibid). Sand was not only easy to outsource but also acted as a medium with sufficient mechanical resistance for roots and plant to grow and, with the correct size of particles, enabled good aeration as well as penetration of water. Generally, these systems did not utilise artificial aeration, which resulted in a lack of oxygen and limited plant growth (Stuart, 1948). Another step forward was attained when, in the New Jersey and Indiana Agricultural Station, sub-irrigation was introduced, with water being pumped into pipes irrigating from the bottom of gravel beds sloping towards the centre of their cross section and towards the end to allow for optimal water flow. The pumping of the water stopped when it nearly filled the channel and the solution flowed by gravity into a tank. In this initial experiment, the system had two tanks, one at each end of the channel (Stuart, 1948).

After the war, in the late 1940s, Purdue University conducted further research and disseminated their results in a series of bulletins in which the soil-less techniques were named *Nutriculture*. There, Robert and Alice Withrow used an inert gravel as the rooting medium instead of sand, with a flood and drain method of irrigation, which allowed plants to be better aerated. This became known as the gravel method (Two Wests, 2019). But although there was commercial interest in such a technique, implementation costs were considered too high. There was no industrialised system available at reasonable costs and greenhouses, possibly the best environment for this new technique of cultivation, were also quite expensive to deploy on a scale large enough to generate sufficient returns to investments. With research focusing on appropriate media, hydroponics became increasingly understood as a soil-less method of cultivation. However, Jones Jr. (2014) maintains that true hydroponics implies that plants are grown without a rooting medium and in a nutrient solution.

The mixture of nutrients dissolved in water was another area of fast advancements. Gericke (1940:17) gave an accurate account of the initial trial and error attempts in this field, noting that, if hydroponics presented a great opportunity to grow food in areas with infertile soil, it also presented considerable challenges since 'the margin of safety against the development of poor growing conditions was much smaller in soil-less crop production'. In the soil, plants seemed to cope better under varied soil conditions and favourable or unfavourable combinations of nutrients available. In water culture, they seemed to be very sensitive to an incorrect proportion of these nutrients. Temperatures would modify the chemistry of nutrient solution much more easily than in soil. In a solution, different nutrients could quickly react to contextual changes and precipitate, becoming unavailable to roots. The lack of a specific nutrient could lead to deficiencies in the growth of plants that may not always be easy to recognise and that would considerably damage crops. Gericke's account outlines the intense period of experimentation in this field that was happening in the US in the first decades of the twentieth century, which led to the modern hydroponic technology as we know it. Gericke also motivated the usefulness of soil-less techniques from a social perspective, pointing out that a large scale hydroponic market could create job opportunities in economically depressed countries with low resource availability and could represent an opportunity to circumvent consequences for agriculture and jobs due to soil infertility, wherever this was a pressing issue (Gericke, 1940:5).

Over the following decades, the diffusion of plastic components made the cost of greenhouses progressively affordable and hydroponic systems increasingly became an attractive option for environments with unfavourable conditions to grow food in soil. The availability of polyethylene films and components such as polycarbonate panelling and plastic drip irrigation tubes led to higher investments in hydroponic systems (Jensen, 1997a, b). However, Jensen (1997a, b) reported, there are limits to the suitability of hydroponics to particular environmental conditions; appropriate locations to install units must be carefully identified in order to avoid excessive costs due to energy consumption for cooling in the summer months. In fact, excessive operational costs did result in financial unviability and the closure of some farms utilising greenhouses. Yet, greenhouse cultivation expanded significantly during the 1950s and 1960s, with large facilities built in the deserts of California, Arizona, Abu Dhabi, and Iran in the 1970s, where solar radiation offered a good environment for maximised plant growth (Fontes, 1973). The trend was reversed following the oil crisis period, started in 1973, which increased the operational cost of greenhouses in terms of energy used for environmental control systems. Progress in soil-less techniques during the 1960s resulted in the development of new technologies with higher efficiency. Eventually, interest in hydroponics crossed the ocean and reached Europe. In the 1970s, the Glasshouse Crops Research Institute at Littlehampton (UK), an agricultural research centre focusing also on glasshouse crops and mushrooms, developed and refined the nutrient film technique which is one of the current options for industrial hydroponic systems (Graves, 1983; The GCRI Trust, 2019).

In the 1980s, the University of Arizona collaborated with Walt Disney Production in order to develop two hydroponic displays at Disney World's EPCOT Centre in Orlando (Hershey, 1994), hosted in the pavilion *The Land* (Roberto, 2000), within which 0.4 hectares were dedicated to experimental horticulture techniques in hydroponics, irrigation methods, and integrated pest management (Bell et al., 2004). Because of its location, the hydroponic unit was used both for entertainment and educational purposes as part of a wider display showing the interaction of men with land. The educational function of hydroponics has been long promoted. Hershey (1994) exposes the advantages of hydroponics in education, allowing the observation of plant roots and the process of growth in absence of 'dirty soil'. The greenhouse hosting the hydroponic units at EPCOT was not only educational and recreational, it was also part of a vast area occupied by greenhouses used to produce the food consumed at EPCOT. Vertical hydroponic towers were utilised, together with aeroponics and special trellises, which, in 2006–2007, yielded 32,000 tomatoes over a 16-month period. At present, 27,000 heads of lettuce per year are harvested. EPCOT is also collaborating with NASA to test hydroponics in space (Farm Flavor, 2019). Experiments in space which focus on food production for self-sufficiency are necessarily using recirculating hydroponic systems as an appropriate technology (Kitaya et al., 2008). In these experiments, the correct choice of crops is vital, not only in terms of growing under particular conditions but also in terms of nutrition. Crops trialled include peanut (Mackowiak et al., 1998), potato (Wheeler et al., 1990) and sweet potato (Kitaya et al., 2008). Experiments on plants in space are not new; however, with the exception of small scale experiments, not much is known about plant physiology in space and how this can be affected in alien conditions. This is a crucial knowledge, enabling survival during long-duration missions and research is being developed in this field (Poulet et al. 2016) as part of future directions for hydroponic systems and their exploitation under a range of adverse environments.

4.1.3 Hydroponic Systems: Functioning and Technologies

This section focuses on the factors and skills which need to be acquired by community farmers or small enterprises in order to build and manage hydroponic systems. The section includes brief descriptions of simplified systems that are accessible to farmers with limited technical and financial resources.

Although plants grown with hydroponic technologies require the same factors for growth as plants grown in soil, these factors are strictly managed by farmers rather than nature when using hydroponics. For example, in soil and outdoors, plants can control their intake of nutrients, absorbing those that are available in the soil in the quantities required. Farmers need to ensure sufficient soil fertility, but each plant acts as any other organism living within an ecosystem and adapts to external conditions for its life. In hydroponic systems, farmers must control many of these vital functions and make up for the absence of soil. The nutrient solution needs to contain an appropriate combination of elements on which the plant feeds and that are necessary to perform physiological functions; water quality needs to be controlled and the correct media must be identified. The following is a list of inputs that farmers must provide and important factors that they need to consider and constantly monitor in hydroponic systems. The list is not exhaustive but provides a good representation of the issues that must be overcome and skills that must be developed by non-experts in soil-less technologies.

4.1.3.1 Nutrient Solution

All plants require a basic range of nutrients to grow but in different quantities and with different additional nutrients. For plant nutrition, 13 elements are essential. These are divided into macronutrients and micronutrients, depending on the quantity needed for correct growth. There are six macronutrients: Nitrogen (N), Phosphorus (P), Potassium (K), Calcium (Ca), Magnesium (Mg), and Sulfur (S). There are seven micronutrients: Boron (B), Chlorine (Cl), Copper (Cu), Iron (Fe), Manganese (Mn), Molybdenum (Mo), and Zinc (Zn). Other elements may be necessary to stimulate growth, but these are not essential or are important for specific plants only. They are called beneficial elements (Rao, 2009) and include Sodium, Silicon, Cobalt, Selenium, Aluminium, and others. Combinations of nutrients are available on the market, with different products depending on the crops. However, many factors can alter the nutrients' availability in the solution, with an impact on the effective growth of crops. Whether the system is recirculating the nutrient solution or not, changes in the chemical composition may occur at any time due to suspended precipitates, microorganisms, organic debris, and other agents. If the nutrient solution is used for an extended period of time, replacement of absorbed elements and filtering of water becomes necessary (Jones Jr., 2014).

Plants with a lack of specific nutrients will show symptoms of such a deficiency, signalling to farmers the need to compensate the nutrient solution with the missing nutrient. For example, iron deficiency typically affects smaller leaves of plants; areas between the veins of these leaves turns yellow while the veins remain green. Deficiency of phosphorus can impede growth, leaves are smaller than normal, and leaf stems and main veins can become reddish-purple (Winterbourne, 2005). For beginners, the difficulty is to develop sufficient experience in order to detect and understand symptoms and, if relevant, understand how to complement existing nutrients' combinations available on the market for specific crops with additional elements. Elements are absorbed in the form of ions with electric charges. A measure of deficiency or excess of nutrient is the electric conductivity (EC) that can be detected in the water around the plant roots. In addition to the observation of plants' symptoms, the measurement of EC can help determine replenishment level required for nutrient solutions.

4.1.3.2 pH

The alkalinity or acidity of the solution is another factor that requires monitoring and adjustments, if necessary. Root activities and nutrient uptake modify the pH, which in turn can have an impact on the correct growth of plants. This changes for each crop with a variation between 5.0 and 7.5. Most nutrient solutions have a pH between 5.8 and 6.5 (Jones Jr., 2014). The pH should be checked every time new nutrient is added to the solution.

4.1.3.3 Oxygen and Temperature

The nutrient solution requires sufficient oxygen levels in order to allow roots to absorb sufficient quantities of water and nutrient. In the soil, oxygen is present because of the air contained in the pockets of a porous medium. In water, sufficient oxygen content must be ensured. In a simplified hydroponic system, this is sometimes obtained by moving the water but most frequently oxygenators can be purchased for this purpose. Temperature and ion composition will influence oxygen levels, thus requiring monitoring. But the temperature can also influence the availability of nutrients in the solution because it can catalyse chemical reactions, thus transforming the composition of elements and their capability of being absorbed by the roots. More generally, any hydroponic system should be designed and managed differently depending on the environmental conditions of the location, indoor temperatures, and availability of natural light. This is not only because of possible overheating of greenhouses or poly-tunnels, but also because of the alteration to the chemical processes that can affect a nutrient's availability.

Other factors that need particular attention include sterilisation and rooting media. Algae and other microorganisms are likely to thrive in water and some of them can be noxious to plants. In order to avoid transmission of these organisms, the equipment needs to be thoroughly washed after every harvest or even sterilised in recirculating systems, using UV lamps. Rooting media can also be colonised by bacteria and algae and will need to be changed or thoroughly washed. Much research has been developed on this area over the last decades (Barrett et al., 2016). Lighter materials compared to sand and gravel, which were used for the initial hydroponic systems, allow lighter vessels and higher operability (Jones Jr., 2014). Popular light media include rockwool, a fibrous material requiring high levels of energy for its production (Rainbow, 2010), expanded clay, or coir fibre, a material that is less energy intensive and completely compostable (Di Lorenzo et al., 2013). This is relevant because locally sourced and sustainable materials are likely to be preferred by environmentally motivated urban farmers. Vermiculate and perlite can be also used, although these too are materials which require high energy use for their production (Rodriguez-Delfin et al., 2017).

4.1.4 Types of Hydroponic Systems

There are several ways to categorise hydroponic systems. Hershey (1994) gave the following broad categorisation: (1) Static, where the nutrient solution does not flow; (2) Flowing, where solution moves continuously through the root zone; and (3) Mist or aeroponics, where roots are intermittently misted with a nutrient solution. It is also possible to distinguish between open (i.e., once the nutrient solution is delivered to the plant roots, it is not reused) or closed systems, depending on whether the nutrient solution is recirculating or lost after irrigation (Jensen, 1997a, b). Today, most of the hydroponic systems are closed loop, recirculating systems (at least those for commercial use), enabling higher water efficiency (Rodriguez-Delfin et al., 2017). Another simple categorisation, reflecting the types that are most commonly used at present, distinguishes plants in a growing medium (i.e., a substrate composed of a solid material in which roots can grow), directly in a nutrient solution, and directly exposed to air in which nutrient solution is periodically nebulised (Rodriguez-Delfin et al., 2017). In these systems, yields are generally higher than with in-soil agriculture due to plant productivity and planting density. However, the main reason for their commercial use is the reduction of soil-borne pathogens and the improved control over water and nutrient supply (Gruda & Tanny, 2014 as cited in Rodriguez-Delfin et al., 2017). The following brief description of hydroponic types firstly introduces the most common ones, which can be used both for medium-to-large commercial farms and small units, and subsequently introduces the simplified types of hydroponics, designed to be affordable and, whenever possible, self-built.

4.1.4.1 Deep Water Culture

This type can be static or flowing, with roots of plants directly in the nutrient solution and plants held above the surface of the solution. Applications of this type vary and can even include simple buckets, in which lids hold the plant and the solution is contained within the bucket, oxygenated with a device. However, it mainly refers to trays designed with the same principles but capable of supporting more than one plant or beds with rafts floating on the surface of the solution. Rafts typically utilise light material such as Styrofoam. Seedlings are grown in blocks (e.g., rockwool or coir) which are subsequently inserted into the rafts in regularly spaced holes. Beds can be of considerable size, therefore appropriate for an industrial hydroponic system. Crops that can be grown with this system are typically leafy vegetables such as lettuce, which are not tall and do not need to be supported with a trellis. Just as in buckets or trays, nutrient solution can be either aerated or recirculated through pumps, therefore enriched in oxygen through movement. The recirculation happens periodically, depending on the size of the bed or the tray.

4.1.4.2 Nutrient Film Technique

This type is a closed system with a flowing solution. Plants are inserted on the lid of a trough or channel, at the bottom of which the solution flows. In this configuration, plants are suspended with roots dipping in the nutrient film. For this to happen, the trough must be set as a slope with a sufficient inclination to move the water by gravity at a recommended flow rate of 0.25 L per minute (Jones Jr., 2014). The continuous flow of solution allows a good aeration of the roots. The water usually flows in a storage below the trough, in which it is filtered before being sent again to the trough with the aid of a pump. Like Deep Water Culture, this type can be built using simple components such as PVC tubes, which are available on the market. Other materials can also be used, provided that the trough is opaque and UV resistant. Commercial systems utilise this type, which is suitable for vertical farming, with troughs deployed vertically either as channels or as stacked shelves, allowing plants to grow on more than one level.

4.1.4.3 Aeroponics

This is the third type of hydroponics which, although very promising for its water efficiency and results, it is not yet as widespread as the other two types. In this type, the plant sits on a surface with the roots exposed underneath to an aerosol of nutrient solution. The advantage is a higher rate of oxygen and lower water and nutrient use. One of the disadvantages is the need for constant monitoring of the system, with loss of power of pumps nebulising nutrient solution likely to cause irreversible damage to plants. Aeroponics was tested with tubers such as potatoes with positive results and presenting very practical advantages (i.e., tubers are completely clean and easy to harvest) (Farran & Mingo-Castel, 2006).

4.1.4.4 Vertical Farming

The definition of vertical farming is still unclear. Many scholars identify it as the hydroponic cultivation of crops in indoor controlled environments, with multiple growing beds deployed vertically on the same surface area. Growing beds stacked vertically can be located in greenhouses or buildings with no solar access. It is mainly associated with an extensive use of state-of-the-art IT, LED, and environmental control technologies, which can maximise the yield and reduce water use and labour, with the typical drawback of being energy intensive and with high initial capital costs (Pinstrup-Andersen, 2018; Nerantzis et al., 2018; Kalantari et al., 2018; Germer et al., 2011). However, vertical farming can be intended as any form of hydroponic vertical organisation of cultivation that can be integrated in and on buildings, including low-cost outdoors systems, which may be suitable in tropical zones, with high solar radiation (Song et al., 2018). One of the possible models of vertical farming uses towers that can host a varying number of plants, drip-irrigated from above, and water collected below and recirculated. Towers are available on the market with different designs. The Zipgrow tower (Zipgrow, 2019), for example, is in metal, with one continuous vertical groove in which plants can be inserted vertically at varying distances. Verti-gro (2019) is a tower composed of segments stacked at fixed intervals and rotated to leave space for plants to grow. These systems of vertical farming can be self-built (see Sect. 4.1.5) and are used in one of our case studies (Huerto Lazo – see Sect. 6.2).

4.1.4.5 Flood and Ebb

This type is quite popular but perhaps not suitable for commercial use. Flow and ebb systems are available on the market. They consist of a plastic tray filled with media such as expanded clay, which has good water retention. Plants in pots are inserted in this medium. The tray is flooded periodically, with the water slowly flowing away to a tank. A pump and a filter are used to recirculate the water. This type is suitable for domestic use, but clay must be washed often to avoid contamination with bacteria, algae, and fungi. Plants grown with this system can be particularly susceptible to root diseases (Jones Jr., 2014).

4.1.5 Types of Simplified Hydroponics

The term simplified hydroponics refers to any soil-less system based on the same principles of conventional hydroponic cultivation but manufactured at a low-cost and easy to operate. Soil-less systems in simplified forms have been promoted by FAO since the 1990s, as they are deemed to be particularly suitable for developing countries. In a technical manual condensing basic scientific knowledge, which is necessary to understand plant development requirements in soil-less conditions, and instructing on the construction of basic hydroponic equipment, FAO presented hydroponics as a system suitable for housewives and children. This is because it can be located in any dwelling's backyard and it does not require heavy labour. The manual was written for Latin America, declaring that simplified hydroponics (which the manual terms as *popular hydroponics*) 'is beginning to become consolidated in the region and in a number of countries is included in national programmes' (Marulanda & Izquierdo, 1993). In fact, in the years following the publication of the manual, simplified hydroponics were installed and tested in several FAO projects across Latin America, including countries such as Argentina, Bolivia, Ecuador, Peru, Uruguay and Venezuela (FAO, 2014). These projects were run within diverse communities, including groups such as those with disabilities and school children, in the recognition that soil-less systems could be used for educational purposes and could be operated by the most vulnerable too. Izquierdo (2005) maintains that simplified hydroponics can generate several benefits which, in addition to those typically attributed to urban agriculture such as community building, include water saving, a lower use of agrochemical fertilisers, and the creation of business opportunities. FAO is still strongly promoting both hydroponics and aquaponics (see following Sect. 4.2) in developing countries, by publishing technical reports with instructions to assemble, for example, simplified, self-built deep water culture systems (FAO, 2015).

Projects in Latin America, deploying diverse types of simplified hydroponics, have been documented in the literature. They all show how such types have been replicated and modified in several ways, adapting to the local availability of components. These projects also show how groups involved in such projects have been successfully trained, demonstrating that specialised skills and knowledge can be learned by all (Fecondini et al., 2009a, b). For example, Michelon et al. (2006) document a project in Brazil, in which a hydroponic unit was built using a wooden container and plastic bottles as water pipes, with the storage tank for the nutrient solution placed above the growing container, refilled manually, therefore with no power input. The water was collected in a second tank at the lower end of the channel and subsequently reused. However, this system required labour; the top tank had to be filled at least four times a day in order to provide sufficient nourishment. Fecondini et al. (2009a, b) document a project in Peru adapting a flow and ebb type of hydroponics, in which plants were inserted in a medium of rice hulls mixed with gravel and contained in a wooden box. The examples above show that the nutrient film technique and other techniques can be applied and modified in line with the concept of simplified hydroponics, often resulting in hybrid types. There are, however, techniques that have been specifically designed to be operated in a simple way.

The World Bank is promoting hydroponics and aquaponics as solutions for food security in refugee camps, especially in water-scarce regions with notable lack of arable land. One of their reports (Verner et al., 2017) maintains that equipment and materials used to build simplified units are available in MENA regions and that the range of hydroponic types ensure that these can be implemented with low initial investment and low technical skills. The report presents as a case study a wick bed system installed successfully in the Palestinian territories and managed by women.

This particular static type comprises a bed with a water tank below and wicks drawing up nutrient solution from the tank to the root zone by capillary action.

Rodriguez-Delfin et al. (2017) mention some types of simplified hydroponics, based on the classic types of hydroponics but modified using easily available components. Deep water culture and nutrient film technique systems can be built out of wooden tanks, PVC tubes, or channels and pumps. A third type is the tower. This type is available on the market in metal or plastic elements, in which plants are inserted vertically, at regular distances, either in holes or in a slit. Seedlings are grown in blocks and subsequently inserted in the towers which are typically drip irrigated from the top. Water can be collected at the bottom of the tower and subsequently recirculated. Industrially produced towers can be quite expensive. One of these towers, for example, is Zipgrow,[1] which is built with durable materials. However, a simplified tower can be made from flowerpots or any other suitable container which is stackable, open at the top and perforated at the bottom, in order to allow pots to be assembled on top of each other whilst being held by a central pole for stability. Pots or containers must be slanted with a base smaller than the top. In this way, plants can grow at the edges of each pot. Media in pots can vary, depending on local availability, and smaller pots containing plants can be inserted in each bigger pot. The irrigation is provided from the top, typically through drip irrigation, and the water collected at the base in a container. Towers can be irrigated manually. They are particularly suitable for plants such as strawberries, with shallow roots and fruit that tolerates hanging. Vertical towers become productively efficient when many of them can be clustered together, allowing high densities of production both on a horizontal surface area and vertically.

A hydroponic type that was designed as a simplified hydroponic system was developed by B.A. Kratky at the University of Hawaii. It does not require power-operated components, water filters, or any other component that is difficult to operate or costly (Atkinson, 2018). A polyethylene-lined tank of appropriate dimension and material (timber, metal or plastic) is filled up to approximately 4 cm with a solution of water and nutrient. A 5 cm deep tray is positioned upside-down at the bottom of the tank, elevating the pots on top of it so as to leave an air gap between pots and the water surface to allow oxygenation of the roots. A mosquito net can be used to prevent insects from reaching the water. Openings at the bottom of the pots will allow roots to reach the water below. The polyethylene membrane covers the tank, with holes allowing plants to grow. Kratky et al. (2005) documented the performance of three variants of the system, each one built with a different container for plants. Recycled aluminium cans, clay pots, or plastic bottles, with aluminium foil lining the bottom and preventing algae growth, were the options experimented. All pots were filled with perlite as a growing medium. The Kratky method is best for small plants. Pots used are typically 10–12 cm tall, with crops like herbs, lettuce, and bushy tomatoes. The system is so simple that it can be installed outdoors, potentially in any condition. Higher yields of tomatoes (2.68 kg/plant of 'Big Beef'

[1] https://zipgrow.com/

tomatoes grown in an aluminium beverage can from a 72-day harvest period) were found to be achievable outdoors, albeit with the system protected from the rain (Kratky et al., 2005).

4.1.6 Productivity

In academic literature, claims of higher yield are quite difficult to verify because they mainly refer to small-scale experiments, rather than data collected from existing farms. Studies also offer insights on a few specific crops only. Jones Jr. (2014) claims that, compared to in-soil production, hydroponics can benefit from a controlled indoor environment and have the potential to have more than one yield per year. In 1930s Gericke quantified hydroponics productivity, measured using the system that he designed (1940). According to his studies, the highest yield of wheat, potato, and rice in the United States was respectively 120; 1150 and 170 bushels (6180; 41,824 and 6182 kg) per acre. Hydroponics can achieve from four to ten times as much. An interesting observation in Gericke's manual is that, with hydroponics, polyculture is possible. Potato and corn, for example, are compatible for multi-cropping because, if mixed appropriately, they would not compete for sun, being of different heights. Typically, polyculture is not practiced in industrial hydroponic farms.

Hoagland and Arnon (1950), the scientists who were asked to assess Gericke's work at the University of California following his refusal to share data from his studies, were more cautious in accepting that hydroponics generate higher yields. They contested that experiments demonstrating such yields are flawed because of the difference in scale between small-scale experiments in hydroponics and extensive in-soil cultivations; higher yields of hydroponic produce were calculated by projecting small scale experiments and comparing them with data of average production in large-scale industrial agriculture. In their publication, they also warned that expectations raised from the experiments in hydroponics and their subsequent media coverage are unrealistic. Claims that 'in the future, most of the food needed by the occupants of a great apartment building may be grown on the roof', and that in large cities "skyscraper farms may supply huge quantities of fresh fruit and vegetables; […] a housewife opening a small closet off the kitchen and picking tomatoes from vines growing in water culture; […] a large chain in New York City is growing vegetables in basements' were untrue and largely misleading (Hoegland & Arnon, 1950:4). The authors seemed to strongly reject all the claims that Gericke made in his manual and through his studies, to the point that, in their writings, they refused to use the term hydroponic (which Gericke coined) in favour of 'nutriculture'. Ironically, the stories that they believed to be mere fantasy have become reality only 6 decades after their publication.

More than 50 years after the publication of Gericke's data on hydroponic productivity, Jensen (1997b) reported the following quantities achieved in a desert greenhouse, comparing it to yields from conventional agriculture. It is important to note

Table 4.1 Yields reported in Jensen's study (1997b)

	Yield/crop (MT/ha)	Number crops/ year	Total yield (MT/ha/ year)	Total yield MT/ha/ year
Crop	Desert greenhouse			Open field agriculture
Cucumber	300	2	600	30
Aubergine	165	2	330	20
Green bell peppers	250	1	250	16
Lettuce	31	10	313	52
Tomato	550	1	550	100

Table 4.2 Yields of lettuce in conventional and hydroponic cultivation as reported in a study by Barbosa et al. (2015)

Production method	Yield (kg/m²/year)	Water use (l/m2/year)	Energy use (kW/kg/year)
Conventional	3.9	250	0.305 (1100 kJ)
Hydroponics	41	20	25 (90,000 kJ)

that the location of the hydroponic farms can have an impact on yields because of temperatures and natural light availability. Table 4.1 reports yields from Jensen's study. Table 4.2 reports yields and resource consumption from another study, comparing soil-less and in-soil cultivations of lettuce (Barbosa et al., 2015). The latter shows considerably higher yields for this single crop when compared to the Jensen study. With unified metrics, the former produces 31.3 kg/m²/year against 41 of the latter. Causes can be either related to the number of harvests per year, which in Jensen's study total 10, or to a higher plant density per m², or advancements in hydroponic technologies and techniques of cultivation.

Finding data on productivity of simplified hydroponics is more challenging. Simplified hydroponics cannot rely on technologically superior equipment, expert advice, and professional experience, possibly resulting in lower yields when compared to in-soil horticulture. Boneta et al. (2019) document an experiment with a simplified open outdoors hydroponic system, located on the rooftop of a building in Barcelona, on an 18 m² surface area. Twenty-two crops were grown in bags filled with perlite with drip irrigation, with a density of 4 plants per m², and a yield of 10.6 kg/m²/year. The most productive crops included tomato, chard, lettuce, pepper and aubergine. In El Alto, Bolivia, the self-built hydroponic unit, located in a small greenhouse, produced 40 kg/m²/year tomatoes (FAO, 2014). This is lower than the yield reported by Jensen (i.e., 55 kg/m²/year), yet remarkable if the inferior quality of equipment and expertise is considered. Clearly, these data need to be considered with caution, but they seem to suggest that non-professional soil-less farmers can achieve good results not only when compared to professional ones but also when compared to in-soil urban farmers. For example, two studies quantifying yields in urban agriculture projects found that, on a sample of 20 sites in Paris and Montreal, these varied between 0.46 kg/m² and 1.96 kg/m² (Pourias et al., 2015); and they varied between 1.99 kg/m^2 and 15.53 kg/m^2 in a sample of 13 gardens in Sydney (McDougall et al., 2019).

Productivity in industrial hydroponic farms is typically not disclosed. In the only case study included in this book, organised at an industrial, albeit small to medium-scale of production (see 6.9.1), directors were in fact reluctant to provide details on productivity, some of which can be read on the enterprise's website and can give a broad idea of the scale of production the case study could attain. For the purposes of this investigation, it was decided to utilise the studies mentioned above as benchmarks because of their contained scale of production and the reliability of data provided.

4.1.7 Policy Context

The EU regulation specifically rules out hydroponics from the organic certification. It specifies that organic certified plants should grow in soil and be fed through the 'soil eco-system and not through soluble fertilizers added to the soil' (Council Regulation 834/2007). 'Therefore hydroponic cultivation, where plants grow with their roots in an inert medium, fed with soluble minerals and nutrients, should not be allowed' in the certification (Commission Regulation 889/2008). A subsequent EU Regulation (2018/848) specifically excludes hydroponics from the category of organic produce. These regulations are a major barrier to the recognition of hydroponics as organic food production, mainly because, in order to be organic, the soil must be the medium for growth. Organic certification recognises the natural ecosystem and its enhancement as the only condition for a horticulture that does not damage the environment. Hydroponic systems, which were born out of the need to find alternatives to a greenhouse horticulture (typically requiring higher fertility inputs and solutions to address the soil-borne pathogens of the plants), do not conform to this principle although are potentially suitable to attain the same aim of the organic certification, i.e., growing produce with a low impact on the environment.

In the USA there are many agencies for organic certification, although they all refer to the National Organic Standards Board. Although the Board is oriented towards denying the certification to hydroponics, a final decision has not been taken. Certification for aeroponics is ruled out. Similarly to the UK, organic certification aims at a restorative use of the soil and organic quality of the inputs, which makes it impossible for soil-less systems to qualify for the certification. However, because of the difficulties in reaching a common view, and until this is reached, the Board concludes ambiguously that hydroponics can be certified as organic if it can be proved that it complies with the prescriptions for the certifications. There is a compelling need to develop a different label for organic certification, which considers the soil-less technology, but there is reluctance in adding another label to the universe of certifications. It is, however, possible to comply with the alternative system of certification 'Naturally Grown', which is peer-reviewed and includes hydroponics (CNG Farming, 2019). Kledal et al. (2019) maintain that there are hydroponic farmers that claim their produce is organic. This adds confusion to a regulatory landscape which is not unified and that, at present, is dynamic and in evolution.

4.1.8 Market Context

The market for hydroponic produce is expanding, although, at present, is quite small. In 2016, in the USA, around 11,000 tonnes of lettuce were produced in greenhouses, 6500 tonnes in vertical hydroponic and 4 million tonnes in field farms (O'Sullivan et al., 2019). Currently, about 3.5% of the worldwide area cultivated under tunnels and greenhouses utilises hydroponic techniques (Sambo et al., 2019). Together with the expansion of the market, research on hydroponics is focusing on the development of ever-more efficient nutrient solutions capable of optimising the intake for each crop as well as the smart management of systems. Sensors and self-correcting algorithm are some of the most promising directions that can advance production. A range of factors can modify the capability of assimilating nutrients or the composition of nutrient solution. For example, the selective removal of nutrients due to plant growth, as well as the process of evapotranspiration, can modify the concentration of nutrients with a negative impact on the growth and quality of the crops. Real time corrections to restore the correct concentration are necessary. To this end, sensor detecting variations and dedicated software correcting the composition of the nutrient solution is only one example of the possibility of the Internet of Things applied to this food technology, which can become part of an approach termed 'Smart Agriculture' (Sambo et al., 2019).

A quantification of the market share of hydroponics is rather difficult. Worldwide, hydroponic production increased from 5000–6000 ha in the 1980s to 20,000–25,000 ha in 2001, reaching 35,000 ha in 2011 (Resh, 2013). Many small and medium-size companies are within this market share although they generate minimal revenue, with only a few large companies generating considerable income. For example, Thanet Earth, the biggest hydroponic farm in the UK, generates USD 121 million per year (D&B, 2020). Reports from market research agencies can help understand the scale of the market, which, as of 2020, is estimated at USD 9.5 billion and predicted to grow to 17.9 billion by 2026. The drivers for growth suggested in these reports include increasing crop land scarcity and low labour operability of hydroponic farms (Markets & Markets, 2020). A report from the UK Climate Change Committee on the UK Carbon Budget for agriculture and land use (2020), recognises that hydroponic technology has the potential to produce 10–50% of the UK's horticultural production. However, it assumes that the real impact on the carbon footprint of agricultural production will come with the soil-less cultivation of crops, such as wheat, with the potential of reducing the land used for intensive production of staple crops. North America is identified as the largest segment of the market, accounting for a share of 35%. Asia and the Pacific region are expected to show the biggest expansion in the near future. The European market will also have a good growth rate thanks to countries such as Netherlands, France, and Spain. It must be noted that this market share also includes companies that produce equipment and components. This is also confirmed by the Markets and Markets report which concludes that the Heat Ventilation Air Conditioning system is estimated to account for the largest market share in the hydroponic equipment market.

4.2 Aquaponics

4.2.1 Introduction

In the literature on aquaponics, historical precedents go back in time as far as the fifth century China and the fourteenth century Aztec empire. However, similar to hydroponics, these precedents refer to initial attempts to farm fish, rather than develop a form of aquaponics as it is currently understood, integrating such a farming within crop cultivation. In these precedents, crops are grown to feed the fish, which are then used as a source of protein by farmers, or fish is farmed and the water of the pond where the fish are is used as the 'nutrient solution' for crops such as rice. In these systems, fish farming and agriculture were organised for human exploitation by intervening and modifying existing ecosystems, rather than – to an extent – creating new ones. In regions with limited exchange of goods with outer regions, resources, including food, had to be found locally as they simply could not be imported from the outside. Building on the tradition of modifying ecosystems for food production and on the scientific advancements outlined in the hydroponics section, a leap forward in the development of a real aquaponic system can be identified in the 1970s, fuelled by increasing concerns about the ecological damage caused by intensive agriculture and farming; the quality of existing food supplies grown in polluted environments; and food security for a growing population. These concerns motivated not only researchers but also groups in society to experiment with food self-sufficiency, the latter embracing a political vision that rejected society and its economic and administrative order. For them, aquaponics was one of the tools enabling independent life, as outlined in the historical section below.

With aquaponics, soil-less techniques move a step forward towards the replication of ecosystems and their closed-loop, nutrient-recovery functioning. Potentially, aquaponics eliminates the need to rely on synthetic nutrients to grow plants (thus being truly low impact) and offers the opportunity to grow within one system many essential nutrients for human diets. In practice, as documented in the case studies, nutrient solution for plants coming from the fish tanks is often complemented with some synthetic micronutrients or macronutrients. This is all the more true for large scale commercial systems. Moreover, like hydroponics, aquaponics often requires high energy inputs. At a small scale, however, this system is promoted as a perfect solution for developing countries. Furthermore, when implemented in community-led/small enterprise projects in the Global North, it can generate many other benefits that go beyond food production. The following sections will outline the history of aquaponics, its current state-of-the-art and advantages when compared to conventional agriculture/farming, and policies available for its uptake.

4.2.2 History

One of the oldest examples of aquaculture and agriculture mentioned in the literature is chinampas. The word is used to describe wetland agriculture in the Basin of Mexico in which long, narrow fields were built above water and separated by canals (Morehart, 2016). This system of agriculture was developed by the local people when they settled in this area, and subsequently deployed and organised at a larger scale between 1150 and 1350 (Espinal & Matulić, 2019), on the lake Texoco, next to the capital of the Aztec empire, Tenochtitlán (today, Mexico City). Since the soil surrounding the city was not sufficiently fertile, the Aztecs turned to a much richer medium: the water of the lake, which was full of nutrients coming from the aquatic organisms. A further advantage was that vegetables grown in chinampas did not have to rely on irrigation or rain but benefitted directly from the water of the lake. Chinampas were built in shallow waters, next to the lake shores, occupying areas up to 330 by 26 ft. (100 by 8 mt). Each area was delimited by stakes and fences below the water level and made from interwoven dead reeds. The resulting space was filled with alternate layers of rock, aquatic vegetation, natural waste and lake bottom soil (Onofre, 2005), containing eutrophic or semi-eutrophic lake sediments (Lennard & Goddek, 2019). Floating rafts built with similar materials and techniques were used as nurseries. These floating and fixed water gardens were deployed to a scale sufficient to provide sustenance to the city and they were governed by dedicated rules and roles. For example, turrets were built overlooking the chinampas, where men could watch the crops and protect them from birds and other animals (Onofre, 2005). The construction of these sub-irrigated, reclaimed gardens required careful design in terms of a substrate with a good capillary fringe and the layer above water of correct depth for the roots to reach the irrigated substrate (Crossley, 2004).

Another often cited precedent for aquaponics is rice-fish culture, which was practiced in Asian countries, including India, Myanmar, Thailand, the Lao PDR, Vietnam, and China. In southern China, fish-rice cultivation techniques were refined and widespread (Halwart & Gupta, 2004). There is archeologic evidence that the qualities of water full of nutrients was well known at least about 1700 years ago, during the mid-Eastern Han Dynasty (25–220 AD). Clay models of a pond with fish inside and a rice field were found in a tomb of that period. Similar clay models were found in other tombs of more recent periods (Renkui et al., 1995). Literature illustrating in detail techniques for rice-fish culture dates back to 889–904 AD, when Liu Xun, in 'Wonders in Southern China', described how carps could live in shallow waters of rice fields, feed off the roots of weed, fertilise the water and ensure the development of rice fields that were de-weeded (Renkui et al., 1995). A more accurate account can be found in a text from the Ming dynasty (about 1573), describing how a pond with fish at the centre of a rice field of several hectares could provide irrigation and nutrient for rice seedlings. Approaches to improve this system were developed until the 1930s, but the introduction of chemical fertilisers interrupted this research and contributed to extinguishing the use of the fish-rice techniques (Renkui et al., 1995).

Recently, the Fujian Provincial Agricultural Science Academy conducted an experiment on rice-duckweed-fish culture. Carps can feed out of the biomass constituted by aquatic plants considered to be weeds, as well as insects such as mosquitoes that thrive in shallow waters. The concentration of nitrogen, phosphorus, and potassium in the water of rice fields with fish culture is higher than without fish. Nutrients contained in the mud is also released through the movement of fish. The experiment demonstrated that, through aquaculture coupled with rice cultivation, ploughing and weeding of the rice paddies can be avoided and crop diseases are reduced. Today, rice-fish culture is common in the Chinese southeastern provinces and in the mountainous areas of the southwestern provinces where fresh fish availability is reduced (Kangmin, 1988). A report from FAO (1957) shows that in the 1950s, rice-fish culture was practiced in 28 countries worldwide, with carp and tilapia as the most popular species used. A more recent report (Halwart & Gupta, 2004) shows that, as of 1999, rice-fish culture is still practiced on over 1200 ha in China, 2900 ha in Thailand, 170 ha in Egypt, 130 ha in Indonesia and tens of hectares in countries such as Vietnam and Madagascar.

Floating gardening was a traditional technique in countries such as Vietnam, Burma, Cambodia, and Bangladesh, where it was utilised for at least two centuries (Islam & Atkins, 2007). In the southern floodplains of Bangladesh, low-income groups could not afford agricultural land and live on flood plains, which were under water during the rainy season. Similar to the Mexican equivalent, they resorted to floating platforms to construct their gardens, using water hyacinth, which was a very common local aquatic plant. The decomposed water hyacinth provides the biomass with which these platforms were built. Above the platform, ash, coconut fibre and, occasionally, soil were layered. The floating platforms were used to cultivate edible plants and seedlings in the rainy season. After the harvest, the platform was dismantled and the residue was used to prepare beds on land for winter crops (Irfanullah et al., 2011). Platforms were approximately 50 × 1.2 m in size, with a depth that ranges from 25 to 50 cm, with approximately two-thirds under water (Islam & Atkins, 2007).

Today, FAO and other NGOs (see for example Practical Action[2] and Welt Hunger Hilfe)[3] promote the diffusion of these ancient techniques as a way for the ever-growing disadvantaged, rural communities to cope with climate change, rising sea levels and flooding events (Saha, 2010). FAO has encouraged aquaculture as a source of protein for the less advantaged (International Center for Living Aquatic Resources Management, 2001), with or without the addition of hydroponics. The assumption behind this strategy is that availability of suitable land and access to

[2] https://practicalaction.org/knowledge-centre/resources/floating-gardens/

[3] https://www.welthungerhilfe.org/our-work/countries/bangladesh/floating-gardens/

protein is limited in many developing countries, especially in an urban context. The limited space required for a small-scale aquaponic unit and the possibility to construct one with low technologies makes this option attractive for households. FAO's technical paper on small-scale aquaponics (Somerville et al., 2014) offers comprehensive instructions covering all phases necessary to implement and run these systems. A 2004 FAO Fisheries Technical Paper too provides specific instructions of the implementation of all these techniques as well as an indication of average yields, gathered in specific case studies. Some of these techniques, however, are only forms of aquaculture combined with other forms of farming (animal-fish systems). For example, the paper illustrates techniques of duck-fish farming and chicken-fish farming in Bangladesh. Because of their behaviour, ducks are particularly suitable for fertilising ponds used as nurseries for graylings with their manure. While swimming, they move the mud at the bottom of ponds, thus releasing more nutrients while oxygenating the water with their movements.

Can these traditional techniques be considered precedents for contemporary aquaponic techniques? They certainly show an understanding of careful management of resource use and closed-loop systems. They also show a propensity to experiment with and combine different food systems in order to have higher control on inputs, outputs, and quantities produced. Traditional techniques constitute a basis on which the contemporary concept of aquaponics was finally developed. However, differently from hydroponics, the history of aquaponics is discontinued between the historical precedents and the recent research leading to the technology as we know it. The New Alchemy Institute is by some authors quoted as one of the initiators of this technology (FAO, 2014; Diver, 2006). The New Alchemy Institute was founded in the early 1970s by John and Nancy Todd in East Falmouth, Massachusetts, as an experiment in ecological self-sufficiency, aimed at demonstrating alternative ways of living that are not environmentally damaging. They were inspired by the Whole Earth Catalogue,[4] a counterculture American magazine offering essays and general information to attain self-sufficiency and light-on-earth lifestyles. The Whole Earth Catalogues were one of the expressions of a generation that was experiencing and reacting to the Vietnam war, race discrimination, a shift in the value system and the psychedelic culture. In this socio-cultural context, John and Nancy Todd conceived the Bioshelter, a dwelling capable of generating electricity and food, the latter achieved through aquaculture (i.e., a pond next to the dwelling), later connected to a greenhouse with a hydroponic system (Todd & Todd, 1976). Their publications documenting the Bioshelter and their experiments with food self-sufficiency helped to promote aquaponic techniques (Diver, 2006). Concomitantly, a team from Southern Illinois University, Carbondale, experimented with and documented the quality of water from a tank with catfish, treated through a biofilter and subsequently used for plant irrigation (Lewis et al., 1978). Finally, in the 1970s Naegel developed the first documented attempt to grow Tilapia fish and Carps in a recirculating aquaculture system, using only the water from the fish tanks

[4] www.wholeearth.com

to grow tomatoes and iceberg lettuce in a Deep Water Culture (see Sect. 4.3) system (Palm et al., 2018).

These initial attempts, as well as a growing preoccupation within society and research groups about the impact of industrial food production on the environment, spearheaded by Carson (2000:1962), were followed by the design, implementation, and monitoring of aquaponic units at the North Carolina State University in the 1980s and the University of the Virgin Islands in the 1980s–1990s, which substantially contributed to establishing an industrial system of aquaponics (Diver, 2006). The North Carolina State University system was designed by McMurray and Sanders. It consisted of a tank of water with tilapia, sunk in the sand floor of a greenhouse. Effluent from the tank irrigated hydroponic containers located on the floor of the greenhouse, using the sand as a medium to grow cucumbers and tomatoes. The system was used to demonstrate the water efficiency of this growing technique and the resulting yields. McMurray and Sanders claimed that water for this system amounted to 1% of that required in pond culture in order to produce equivalent tilapia yields. Another notable feature that was noted in this system was the role of sand and vegetable roots as proper filters, capable of restoring the water quality of the effluent water (reciprocating filters). Finally, McMurray and Sanders also researched into the ratio of water to growing medium which, with its variation, can favour either higher fish or vegetable yields (Diver, 2006). In parallel with the academic investigation in the early 1990s, the company S&S Aquafarm in Missouri modified the North Carolina State University aquaponic unit and developed a full-scale commercial unit. One of the main modifications from the original system was the use of gravel instead of sand for the vegetable beds (Datta, 2015). The design of this commercial unit was called Speraneo, which is the name of the couple who owned the company. This design became very successful and was widely replicated in other commercial units. S&S Aquafarm could produce between 45 and 70 pounds (20.4–70.7 kg) of vegetables – calculated as cumulative yield in the period of the fish development – for every pound of tilapia (0.45 kg). The Speraneo aquaponic method demonstrated one of the most important characteristics of aquaponic systems: the generation of higher yields of vegetables compared to fish (Diver, 2006).

The system developed at the University of the Virgin Islands by a team led by Rakocy further advanced the aquaponic technology, establishing correct volumes, procedures, and sequences of fish-plant cycles. The team built a commercial scale aquaponics facility with four 7800 L fish tanks with nile and red tilapia, and a total of 240 m^2 hydroponic troughs with basil, lettuce, okra, and other crops. Calcium hydroxide and potassium hydroxide were regularly added to the water to maintain an acceptable Ph (7–7.5) and iron was added as a necessary supplement for plant growth. Tilapia was grown for 24 weeks, and the harvesting was alternated in a way that a tank was harvested every 6 weeks. Rakocy also documented annual quantities of fish and basil (one of the main crops produced), which amounted to a value of USD 134,245 (Rakocy et al., 2003).

Today, studies on aquaponics enable a better understanding of the conditions necessary for this technology. Recently, the COST Action (a European fund to start and establish networks of researchers and professionals on specific topics) on aquaponics, developed a new definition of aquaponics, reflecting such progress. In fact, the original definition formulated by Rakocy et al. (2003) is: 'aquaponics is the combined culture of fish and plants in closed recirculating systems'. The new one from this COST Action is more sophisticated and requires that aquaponic systems farm sufficient fish to generate the nutrient required to feed the plants grown hydroponically (at least 50%) in the same recirculating system. In fact, systems containing a very low number of fish when compared to plants do not effectively recover and use nutrient within a recirculating system. It must be noted that this definition does not reflect – yet – what happens in industry, where a vast variety of aquaponic system types can be found (Lennard & Goddek, 2019).

4.2.3 Aquaponic Systems: Functioning and Technologies

Aquaponic technology merges hydroponic and aquaponic technologies. Experiments with hydroponics were initiated in order to address issues of soil fertility and soil-borne pathogens, which are those that typically must be addressed in indoor horticulture. Experiments with water recirculation were part of attempts to optimise the use of water to rear fish (aquaculture) using filters, which were particularly relevant to water scarce environments. These experiments used domestic wastewater treatment technologies and were aimed at developing a system in which water is filtered and recirculated, and stabilised with oxygen and carbon dioxide contents. The resulting system is named 'RAS' (Recirculating Aquaculture System). Although experiments started in the 1950s, it took more than 40 years to refine and establish such systems. Today, several fish species are grown with RASs, and they are also increasingly utilised to produce larvae and fingerlings (Espinal & Matulić, 2019). RASs are basically composed of three elements: a device to filter and remove solid particles from the water; a biofilter capable of processing the ammonia excreted by fish into less harmful nitrates; and a device to extract carbon dioxide from the water while enriching it with oxygen (Espinal & Matulić, 2019).

The idea of coupling these two technologies comes with several advantages and challenges. The advantage of a closed loop system in terms of water savings and maximisation of nutrients is evident. Water replenishment requires, at least in theory, the same quantity that is lost through the process of plant evapotranspiration and sometimes evaporation. Fish feed is another external input which is, however, well utilised, since it is transformed by the fish into nutrient for plants. Energy is the third input, used in relatively small quantities to power pumps, but in very large quantities, although not always, for indoor hydroponic units utilising artificial light and environmental control systems. Water replenishment can be minimised with good management of the system and electricity can be produced with on-site renewable energy devices. This leaves only fish feed as an external, substantial input. Some on-going experiments are attempting to address this issue too by growing in situ larvae as fish feed.

Practical challenges include the fact that plants and living stocks require different types of nutrients. Plants need elements that fish cannot provide. For example, iron and other elements, may not be supplied through fish excretion (Goddek et al., 2016). The management of aquaponic systems is quite complex in terms of matching the number of fish and the quantity of the fish feed (i.e., concentration of nutrient in water) with the number of plants and their nutritional requirements. Rakocy (2012) gives some general indications to attain this balance (e.g., 57 g of feed/day per square meter of plant growing). Likewise, plant retention of nutrients that can be harmful to fish must be monitored. This requires a correct selection of plants and fish. For example, the Nile tilapia can withstand higher nutrient load (Palm et al., 2019). Generally, a correct combination of plants can result in higher yield of vegetables and even in higher fish welfare, since some substances released by plants are beneficial to fish health (Palm et al., 2019). Finally, it is important to avoid contact between fish excretion and edible plants, which requires the correct design of the system and procedures to grow and harvest. All these challenges as well as others that are well documented in the literature, require expertise in design and management of the aquaponic system or at least perseverance in testing the system until a balance is reached. As we will see in the case studies, the size of the aquaponic systems in most community-led projects is small and units are rather simple in terms of components and control systems. Choices of crops and fish species are often not determined by efficiency of the overall system but by chance (a fish species that is available locally and inexpensive) or curiosity. Yet, even constructing and running these simple systems requires research and a process of trial and error that would not be necessary if a community-led urban agriculture project were utilising conventional horticultural techniques.

Backyard Aquaponics is a broad term used internationally to describe any low-cost, self-built aquaponic that can be placed in people's backyards and used for personal consumption. The Backyard aquaponics website[5] offers resources, a forum and downloadable material for beginners. The website is not linked to any organisation with a clear programme aimed at promoting food self-supply but rather is a bottom-up, global initiative. It is not the only website of this kind (see for example Friendly Aquaponics).[6] Backyard, simplified aquaponics has been promoted for a long time by FAO and embedded in national programs in developed countries. A backyard aquaponics program was organised between 2010 and 2016 for local communities in Hawaii, where about 85% of food is imported. More than 80 households were recruited and participated in training workshops (Beebe et al., 2020). A similar initiative was organised in Indonesia, leading to more than 80 people forming an aquaponics community (Rahdriawan & Arriani, 2020).

Many of the elements composing an aquaponic system are the same as those described in the Hydroponic section. What follows is a description of the components that must be added to a hydroponic unit to become a typical aquaponic unit.

---

[5] http://www.backyardaquaponics.com/

[6] https://www.friendlyaquaponics.com/backyard-aquaponics-systems/

Skills and knowledge necessary to assemble components and monitor the resulting equipment, in addition to those noted for hydroponics, are different. Perhaps the biggest challenge is the farming of the fish, which necessitates an understanding of its growth cycle and wellbeing.

4.2.3.1 Filters

This is a fundamental component of the system; fish feed and excrement need to be filtered in order to maintain a sufficiently good water quality. Water with an excess of solid waste increases the risk of fish diseases, increases the level of ammonia, and decreases oxygen concentration. Filters must be positioned between the fish tank and the biofilter. There are many options available for filtering. These include filtration by gravity (sedimentation, swirl separators/radial flow separators), screens, or more sophisticated options (e.g., oxidation through ozone treatment). Filtering by gravity does not require mechanical, energy operated equipment and it is therefore more affordable and appropriate for small scale units. High volume of water will need filtering through screens.

4.2.3.2 Biofilters

This is another fundamental component, especially considering that the conversion of ammonia into nitrate – which plants can process – is at the core of the recirculating water system. A biofilter is a tank containing a porous media on which bacteria naturally occurring in the environment can grow. It is advisable to monitor levels of ammonia nitrite and nitrate in the water contained in the biofilter in order to ensure that a sufficiently large population of bacteria is in the biofilter. New biofilters can take as long as 6 weeks to effectively function. Once ammonia is converted into nitrate, this is no longer harmful for fish and can be absorbed by plants as a source of nitrogen.

4.2.3.3 Temperature and Other Factors

Optimal water temperature for fish tanks varies, depending on the fish species. Cold water fish live in temperatures between 10 and 18 °C. Lower temperatures can increase the risk of diseases. It is possible to bring the water to optimal temperature, when necessary, but this requires additional energy, thermal insulation of the tanks, suitable equipment, and consequently higher costs. A wiser approach would be to farm fish that can live well within local environmental conditions. Oxygen levels are also essential to the survival of the fish. In fact, oxygen levels are responsible for many fish deaths in aquaculture (Lennard, 2012). High water temperature decreases the solubility of oxygen. Dissolved oxygen (DO) should be maintained at least above 5 mg/L. The sizing of the system is another factor that is quite difficult to attain. For example, with the hydroponic component undersized, nutrients dissolved in water will start accumulating because they are not absorbed in sufficient quantities by the plants. Conversely, there may not be sufficient fish to generate the quantity of nutrient necessary to feed the hydroponic unit. General ratios are available in manuals, but a balance will differ depending on crops, fish, dimension of the unit, and more. It is therefore important to monitor the system to ensure its correct design. Connected to this is the fish feed rate, which determines the quantity of food which will generate enough nutrient. The FAO simplified aquaponics manual suggests 40–50 g/m^2/day for leafy vegetables and 50–80 g/m^2/day for fruit.

4.2.3.4 Animal Wellbeing

As noted, one the challenges for aquaponic farmers is fish wellbeing. Diseases derived by fish pathogens can be worsened by factors such as the wrong nutrition and an adverse environments. Both can generate stress in fish and result in economic loss for farmers, whenever infections spread. But fish health can also be affected by abiotic factors, such as low oxygen content in water, high ammonia content, and water temperature. Strict sanitation and water content monitoring are important preventative measures, together with avoidance of overstocking and sufficient knowledge of each fish specific disease (Yildiz et al., 2019).

4.2.4 Coupled Aquaponic Systems

In a coupled aquaponic system, aquaculture and hydroponics are collaborating to form one system in which water is shared and recirculated. The aquaculture components comprise a water tank (or tanks) with the fish. This water is filtered in a clarifier in order to eliminate solid particles such as the fish excrements and uneaten fish feed. This solid waste is collected and sometimes processed to be used as fertiliser. The water is subsequently transferred to a 'biofilter', a tank filled with substrate for the growth of bacteria, where the ammonia nitrogen contained in the clarified water is processed by bacteria through oxidation and transformed into nitrates (nitrification), which plants can utilise as nutrients. Substrate in the biofilter is usually made of plastic or ceramic small elements which accumulate bacteria on their surfaces. Biofilters were not used in initial experiments with aquaponic units, which relied on the hydroponic system substrate (often using gravel) for nitrification. Aquaponic units can function without biofilters although these are always included in medium to large professional units, designed for intensive production. The water is subsequently, but not always, enriched with nutrients that are necessary to plant growth.

This last point is contentious: as noted aquaponics can potentially overcome the use of synthetic fertilisers, which could reduce its carbon footprint and, to an extent, justify the claim that hydroponic produce can be organic. In one of the case studies, there is no use of synthetic nutrients (Bioaqua Farm – see Sect. 6.6). Lennard (2012) maintains that, today, it is possible to provide 90% of plant nutrient through fish feed, provided that the solid waste from fish tanks is processed via a biodigester and reintroduced in the recirculating system. This option would minimise substantially the input of synthetic nutrient even when this is deemed necessary.

There is no specific requirement for the choice of the hydroponic technique and ebb and flow tanks, nutrient film technique or deep water culture can be all utilised. The water that comes out of the hydroponic trays or tanks, cleared of the nutrient at an acceptable level, is then recirculated in the fish tanks with the aid of a pump. Another component that is added in professional aquaponic units is the purifier. Water can contain pathogens that are harmful to fish. Purifiers using UV light, which alters the DNA of bacteria, thus preventing their reproduction, can sterilise the water. According to Lennard and Goddek (2019), currently there is a tendency to avoid sterilisation because it can inhibit the growth of a bacterial fauna which is beneficial to the entire system. Without it, the system reaches an overall ecological balance and – it was noticed – this enables plants to thrive.

The several variables outlined in this brief, general description of a coupled aquaponic system allow many possible configurations, with several levels of sophistication and many different designs. It is possible to include or exclude some components and test different techniques; or to play with the concentration of fish and the composition of the fish feed; or to test the coupling of several fish species and match them with a selection of crops to enhance fish welfare or plant yield and health. It is also possible to vary the size of the aquaculture component in relation to the hydroponic component. A water tank with a small number of fish, not sufficient to provide nutrient to the hydroponic unit, can also be an option (although perhaps not in line with the definition of aquaponics as formulated by the COST Action Aquaponics) because it is always possible to add nutrient to the water before it enters the hydroponic component.

4.2.5 Decoupled Aquaponic Systems

Aquaponics is a technology in development. One variation which has recently been considered as an advancement is decoupled systems. In coupled systems trade-offs are necessary in order to attain a quality of water that is sufficiently good for fish and plants, which would normally require different water conditions. This can be a major drawback, influencing, for example, the choice of crops and fish species. In a decoupled system, the nutrient solution does not recirculate, but rather is kept in two different loops composing the system. This approach enables a higher control on the quality of water and the nutrient mix, which can be calibrated independently for plants and fish. For example, the pH of the two systems can be better controlled. The multi-loop design of decoupled systems provides for a process of anaerobic digestion of the solid waste from the aquaculture component and of biomass from the crops (Goddek et al., 2019a, b). Decoupled systems require a higher specialist understanding of the aquaponic cycles and are perhaps less likely to be used in community or educational projects, as we will see in the case studies.

Other more experimental attempts are being tested. These include the merging of aquaculture and aeroponics (currently not tested yet); vertical aquaponics (the merging of aquaculture and vertical farming, leading to a more intensive use of the space for crops) and maroponics, in which saline water is utilised, with appropriate fish species and plants that grow with a low-to medium salt concentration in water (i.e., brackish water), such as samphire (Kotzen et al., 2019). Unlike hydroponics, there is no specific technique for simplified aquaponics, or backyard aquaponics. The simplified hydroponic techniques can still be used in aquaponic systems, and simple tanks built with Intermediate Bulk Container (IBC – plastic containers with a metal cage, used for truck transport or boat transport) are connected to a filter, a biofilter and the simplified hydroponic unit.

4.2.6 Productivity

Generally, compared to hydroponics, good levels of crop harvest can be difficult to attain in coupled aquaponic systems. This is because in hydroponic systems, nutrient solution can be calibrated to species of plants cultivated. In coupled aquaponic systems, water is both the nutrient solution for plants and the medium for fish. Quantities of nutrients added to the water coming from the fish tanks, which are deemed necessary for plant nutrition, must be exactly those that plants can absorb, since fish may suffer from excessive concentration of these nutrients in the water. Also, pH requirements for fish and plants are different. However, since fish are organisms with stricter requirements than plants, it is critical that water quality meets the requirements of the former (Lennard & Goddek, 2019) and this may hinder productivity of plants when compared to a hydroponic system (Kledal et al., 2019). However, Lennard (2005) experimented with productivity of lettuce in aquaponics and hydroponics, obtaining similar yields after several trials and optimisations of the aquaponic system utilised (5.77 kg/m^2 of the aquaponic lettuce, compared to 5.46 kg/m^2 of the hydroponic one).

Productivity in aquaponic systems is very difficult to establish because each system can be designed differently, with density, choice of crops and fish species, and balance between the size of aquaculture and hydroponic units being only some of the factors that can influence it. Love et al. (2015a, b) recorded the productivity of a 10.3 m^3 aquaponic system located in a 116 m^2 poly-tunnel in Baltimore. In 2013, they recoded a total of 129 kg of fish, not all of which were harvested, and 294 kg of lettuce, kale and herbs. The following year, they measured 117 kg of fish and 422 kg of the same crops. The feed conversion ratio (the weight of feed intake divided by weight gained) over the 2 years averaged 1.29.

Table 4.3 below is based on a study by Bailey and Ferrarezi (2017), providing benchmarks for the commercial value of several crops. Yields on which this study is developed are those from the Rakocy's aquaponic unit at the University Virgin Island, built with outdoor hydroponic beds, in a climate allowing several harvests per year. It is worth considering that the number of harvests per year could be higher in an aquaponic indoor farm with a controlled environment (see also Rakocy et al., 1997).

Other experiments recorded different yields, demonstrating how varied productivity can be, being sensitive to parameters such as size of the aquaponic unit, climate, composition of the nutrient solution and more. In an aquaponic unit built in three greenhouses in the Crop Diversification Centre South Brooks, Canada, based on the design of the Virgin Islands unit, Savidov and Brooks (2004) recorded a yearly yield of tomatoes and aubergines, respectively 20 and 9 kg/m^2, together with exceptional yields of 51.5 and 58.3 kg/m^2 of swiss chards and water spinach. In a study on three aquaponic farms in Hawaii, Tokunaga et al. (2013) ascertained the productive capacity summarised in Table 4.4. The average ratio between fish tank volume and vegetables grown is 1.47 gal./sq. ft.

Productivity in simplified aquaponic projects is equally difficult to identify. FAO defines simplified aquaponic projects (termed in their report 'small-scale aquaponics'), units with a fish tank size of about 1000 litres and growing space of about 3 m^2 (Somerville et al., 2014). This definition is reductive; case studies with simplified aquaponics that are documented in this book have slightly higher capacity and yet are self-built and not commercial. The same report provides the following average yearly yield that can be expected from small scale units, for two of the crops cultivated and the fish. Yields provided are higher when compared with the University of Virgin Islands yield (Table 4.5).

Table 4.3 Yields of Rakocy's aquaponic farm at the University of Virgin Islands (Bailley & Ferrarezi, 2017)

	Plants/m^2	Harvest per year	Yield kg/m^2	Yield – total Kg/m^3/year
Lettuce	16	12	27.84–50.88[a]	
Pak-choi	30	4/5	8	
Kale	30	4/5	0.89	
Swiss chard	30	4/5	1.44	
Cucumber	8		6.2	
Okra	4		3.04	
Zucchini	2.7	8	7.6	
Tilapia				16.02[b]

[a]Lettuce harvest is expressed in heads. Conversion to kg is based on Rakocy et al. (2011) (1 Romaine Lettuce = 0.265 kg) and Barbosa et al. (2015) (1 lettuce plant = 0.145 kg)

[b]The fish yield is based on a capacity of the fish tanks of 31.2 m^3. However, the total water capacity of the systems is 110 m^3, including filters, clarifier and hydroponic troughs. The yield is calculated on the fish tanks capacity only (Rakocy et al. (2011)

Table 4.4 Yields of three different farms as reported by Tokunaga et al. (2013)

	Farm A	Farm B	Farm C
Total growing area	1070 m^2	1141 m^2	2650 m^2
Tanks - total litres	68,000	27,000	340,000
Total vegetables sold	83,102 kg/year Lettuce	28,304 kg/year Lettuce	30,037 kg/year Lettuce, tomato, cucumbers and beets
Total fish sold (unit?)	567	n/a	7095

Table 4.5 Yields of a backyard aquaponic unit as reported by FAO (2014)

	Yield
Lettuce kg (no. head x weight per head)	(360 × 0.18) 64.8[a]
Tomato kg	54
Fish kg	30

[a]Lettuce harvest is expressed in heads and kg, using as conversion Tokunaga et al. (2013) (1 lettuce plant = 0.18 kg)

4.2.7 Policy Context

As for hydroponics, in Europe there is no organic certification available for aquaponics. Fish sold as organic need to be bred in open organic pond systems (Thorarinsdottir et al., 2015; see Council Regulation 834/2007; Council Regulation 889/2008, Commission Regulation 710/2009). Also, aquaponics often necessitates synthetic inputs that are not permitted for organic farming. Kledal et al. (2019) maintain that a certification for ‘organic’ aquaponics is timely and necessary because it would justify a premium value for food harvested with this technique, compensating for the high capital investment that the equipment requires. This certification would also recognise and reward the resource efficiency – therefore embedded sustainability - of this technique. As noted, the soil-less condition for plant growth is incompatible with EU regulations for organic horticulture, which specify soil as the only acceptable media for plants, with no synthetic inputs allowed.

The EU regulation recognises the resource efficiency of aquaponic systems but rules them out from organic certification on the basis that they stray away from the natural medium where fish typically grow, although it allows the production of fingerlings. The regulation states that ‘due to the principle that organic production should be as close as possible to nature, the use of such systems should not be allowed for organic production until further knowledge is available. Exceptional use should be possible only for the specific production of hatcheries and nurseries’ (Council Regulation 889/2008; Commission Regulation 710/2009; para. 11). Fish stocking density in tanks is also a problem since in farmed fish, densities are typically lower than in tanks for aquaponic systems. The USA context for hydroponics described above is valid for this technology too, with the further complication that there are strict standards issued by the United Stated Department of Agriculture to treat animal manure in organic production and none of these standards refer to fish manure for plant production. In the USA, standards and certifications for organic fish are available. Some of them can be found in Norms for Organic Production and Processing (NOP). But the NOP does not clearly define aquaculture, and this has been a barrier to the growth of this sector (Diver, 2006).

Other issues limiting the diffusion of aquaponics in Europe include the lack of a clear classification for aquaponics; that is, animal and plant production together, which is not recognised in any policy relevant to aquaculture or agriculture. This limits the eligibility of enterprises in this sector to be recipient of funding under the Common Agricultural Policy. In addition, this policy does not recognise urban agriculture as a category for food production, therefore funding for enterprises operating in cities is not available (Hoevenaars et al., 2018). Yet, aquaponics, together with 3D printing, drones and wearable technologies is included in a report commissioned by the Scientific Foresight Unit of the European Parliamentary Research Service, which identifies some of the innovative technologies that, if scaled up, could potentially improve lives and generate a positive impact on society. The same report recognises that because of the cost and the limited range of fish species and plants which can be grown (at least with the current state-of-the-art techniques), aquaponics is well suited for urban environments, enabling zero miles, low-input food and a certain degree of self-sufficiency. To this end, the integration of these systems in buildings should be developed to a further extent (Van Woensel & Archer, 2015). Subsides are key to the survival of agricultural and farming activities which are not competitive at a global level and therefore not commercially viable for some countries in the current market system. To date, funding opportunities for research on aquaponics are available through funding programmes such as Horizon 2020, but there is little funding available for enterprises to start and develop their businesses. There are environmental constraints too, related to the legislation that prescribes separation of the effluents of aquaponics for plants and for animal production. Generally, in EU member states, solid fish excrements are currently considered fish waste and not plant nutrient (Miličić et al., 2017). In Germany, with a regulation that proves tighter than the EU policy, hydroxylates of animal origin are not allowed either, questioning the suitability of the fish-derived nutrient for plants grown with aquaponics (Kledal et al., 2019). Yet, EU regulation recognises the organic certifications of plants that are grown on soil, allowing a certain percentage of fish waste to be used as nutrient, similar to manure in organic horticulture.

Today, research on aquaponics is no longer predominantly developed in the USA, as during the last decades, but also elsewhere. The role of FAO in promoting this technology in developing countries and a strong focus by the European Commission on research on sustainable food production and rationalisation of resources has attracted researchers from other countries. As a spin-off of the COST Action on Aquaponics, an EU Aquaponics Hub[7] was established. European research funding programmes such as the Seventh Framework Programme and Horizon 2020

[7] www.euaquaponicshub.com

have also funded some projects on aquaponics such as EASY, CoolFarm and ECOFISH.

Hoevenaars et al. (2018) maintain that, although there is no specific mention of aquaponics and hydroponics in EU agricultural policies, the specific focus on the regulation of effluents in aquaculture and increased competitiveness through innovation can be a driver for enterprises to invest in this technology. Access to space and water for aquaculture can be problematic and competition between countries can limit the development of this sector. Aquaponics offers a solution both for space competition and water pollution. On the other hand, aquaponics combines agriculture and aquaculture, whereas European legislation deals with each sector separately. There are only a few commercial aquaponic farms operating in Europe. A recent report lists 10 such farms either working or under construction in Europe and several hundred large farms (>100 m^2) operating worldwide (Villarroel et al., 2016). The European aquaponic farms include research facilities and enterprises such as Breen, a SME with a 500 m^2 facility in Hondarribia, Spain, which has been operating for many years; Institute of Global Food and Farming (IGFF) in Denmark, with a decoupled aquaponics unit of 60 m^2; the Icelandic company Svinna-verkfraedi Ltd., which, in collaboration with the University of Iceland, implemented an RAS system with tilapia in the greenhouses of Akur, an organic greenhouse horticulture farm in South Iceland; Nibio, in Grimstad South Norway, which is a facility used for testing techniques and optimising processes; the SME Ponika in Slovenia with a 400 m^2 commercial aquaponic system; Eureka Farming, Italy, with a 500 m2 experimental facility in which fresh and sea water fish has been successfully reared; and Urban Farmers, a rooftop research facility in Basel (Thorarinsdottir et al., 2015). However, in a 2016 study, Villaroel et al. (2016) estimated that, in Europe, there are approximately 20 large aquaponic farms (above 1500 m^2).

4.2.8 Market Context

The farmed fish market is sizeable. In comparison, that of aquaponic farms is almost negligible. According to FAO (2020a, b), in 2018, fish farmed both in marine environments and inland comprised about 46% of the global catch, corresponding to 82.1 million tonnes out of 178.5 million tonnes. A report by the University of Sterling and RAS Aquaculture Research Ltd. (Murray et al., 2014) maintains that current aquaponic fish production is less that 100 tonnes, about 0.0006% of the global aquaculture market. Yet, aquaponics is growing. In 2002, Jones (2002) reported that 'there are less than five large aquaponic farms worldwide (around 1 acre) and many other small and family farms'. A few years later, in 2015, a publication with Aquaponics Guidelines, co-funded by the Eco-Innovation Initiative and the European Union (Thórarinsdóttir et al., 2015), stated that a few large aquaponic units, as mentioned above, are being built or already established in European countries, with many small-scale farms being built as experimental units, initiated as part of research projects or collaborations between research institutes and industry. A survey organised with a sample of 68 aquaponic farms showed that the majority were established as research facilities or demonstrators. For an overwhelming share of the sample (above 80%), aquaponics was not the main source of income; 46 produced less than 100 kg and only 12 produced more than 1000 kg fish and plant per year (Villaroel et al., 2016).

The situation is different in countries such as the USA and Canada. A report from CISION, a market research organisation (Cision, 2020) estimates that, as of 2020, the USA aquaponic market is worth $630 million, and represents 42–45% of the global market. Other market research reports stress that aquaponics is still a small market in the USA, although highly promising, and that in recent years many new enterprises were started. Mordor Intelligence website states: 'In Wisconsin, the number of aquaculture farms recently rose from 2300 to 2800, with 300 out of the 500 new farms being aquaponic farms' (Mordor Intelligence, 2020). Although studies suggest that aquaponic systems can be very productive, the economic viability of an aquaponic enterprise can be problematic. An international survey was conducted by Love et al. (2015a, b), with 257 respondents, including farms and suppliers of aquaponic equipment. Most of the respondents were small businesses (by size of revenue i.e., less than $50,000 per annum). Only 31% had been profitable in the previous year and those who sold equipment – thus not only operating as a farm – were the most commercially successful. Moreover, businesses selling only crops were more positive about their future, thus suggesting that the market for fish is more competitive and that crop sales yields higher monetary returns.

Beyond policy constraints that may limit the uptake of aquaponics (e.g., incentives and the lack of certification for organic aquaponics), there is also the issue of acceptance of fish and hydroponic produce, which may not be perceived as desirable. A study documenting a survey in Berlin by Specht and Sanyé-Mengual (2017) showed that only 28% of interviewees viewed aquaponic production in cities positively and only 27% expressed willingness to buy aquaponic products. Another survey by Miličić et al. (2017), showed that of over 635 respondents to an online questionnaire, more than 50% had no prior knowledge of aquaponic technology. Yet, fish available to consumers is largely farmed and sold as such, and the impact on the environment of farming practices is well documented. This impact is connected mainly to food that is not eaten by the fish and excretions that accumulate on the sea floor underneath cages, leading to low-oxygen conditions (Rubio-Portillo et al., 2019). Organic fish farming techniques are less harmful but still not established within the farmed fish market (Di Marco et al., 2017). Against this backdrop, aquaponics – and urban aquaponics too – is an opportunity to produce low-impact food. An effective information strategy communicating the value of such an opportunity can lead to greater trust in soil-less and aquaponic food technologies. We are still in a transition phase in which uncertainties over the uptake of this technology are high. Over the time this book was drafted, several farms were opened in the EU, and others closed (e.g., Urban Farmers in Den Haag). This is a sign of how difficult it is for a new food technology to succeed within the current food systems.

4.3 Mushroom Farming

4.3.1 Background and History

The third soil-less farming technology included in this book, mushroom farming, differs substantially from the other two in approach to cultivation and industry characteristics. Mushroom farming has a long-standing tradition, which has evolved as an established and still expanding industrial activity, whereas hydroponic and aquaponic industry is still at its infancy. The global mushroom supply comes predominantly from commercial farming, although foraging wild mushrooms is a thriving industry that can generate considerable income for communities living in regions with forests where mushrooms can be harvested (Tsing, 2015; Zhang et al., 2014; Yang et al., 2008). Large scale farming relies on expensive equipment and technologies, but the cycle of mushroom growing can be carried out with low-cost tools and materials, therefore being suitable for small farmers across the world. Yet, to our knowledge, over the last decade, this soil-less cultivation technology has attracted less interest than hydroponics and aquaponics in community-led urban agriculture projects. There are two urban mushroom farms documented in this book, but their agenda is not so infused with social values as the other case studies. The higher uptake of hydroponics and aquaponics in community projects is perhaps a consequence of their novelty and the perceived innovative character of these technologies. Another hindering factor may be related to the fact that mushroom farming is very much specialised, whereas the other two technologies allow the cultivation of a wide range of crops. Yet, mushroom farming is a particularly interesting technology because, although indoor environmental control is necessary to attain optimal conditions for growth, mushrooms can still be grown using substrates that are natural, with very little waste and replicating the actual conditions necessary for them to grow outdoors. For the reasons mentioned, the sections on the history, technology, and techniques of mushroom farming is less developed than the previous two sections, although equally relevant for mapping trends in soil-less urban food production.

Mushrooms are fungi that developed from algae (Stamets & Chilton, 1983). There are about 12,000 species of fungi that can be defined as mushrooms, 2000 of which are – to varying degrees – edible (Sanchez, 2004). A mushroom is the 'fruiting body' sprouting from the mycelium, a perennial network of cells that grows underground and that, under ideal environmental conditions, produces the fruiting body that we eat. The mushroom plant germinates from spores, which are generally stored on the gills under the cap of each fruiting body (Stamets & Chilton, 1983). Mushrooms cannot feed through photosynthesis as plants. Instead, they use their enzymes to break down compounds such as cellulose and lignin, to subsequently absorb the degraded compounds through the hyphae, the thread-like elements composing the mycelium. In nature, many substrates contain such compounds and depending on the type of mushroom, mycelia will grow on logs, beds of straws or even manure.

Substrates can be easily reproduced for farming, and they can be built with by-products of agricultural production, woodland, animal husbandry, and other industries (Rinker, 2002), making mushroom farming suitable for regions with high availability of agricultural waste (Higgins et al., 2017) and particularly suitable to practice as a form of agriculture based on circular economy principles (Dorr et al., 2020). Once used, spent mushroom substrate can be turned into animal feed or soil enhancement (Sanchez, 2004). Mushrooms can be used for many purposes, including in the pharmaceutical industry. Because of their capability of processing organic compounds, they can be also used for bioremediation (Chang, 2009). Although with a low-calorie and carbohydrate content, mushrooms are an excellent source of protein, containing all the amino acids necessary for human diets. In fact, the amount of crude protein in mushrooms is higher than milk. They also contain important nutrients such as phosphorus, iron, and vitamin B (Chang, 2009), therefore representing a food with high nutritional value. Mushrooms are also a potential substitute to meat. Recently, new animal-based foods have been experimented with where mushrooms were blended with meat, thus reducing the meat content by 25% (Lang, 2020a, b). Because of their properties, they enrich many other processed foods such as bread, pasta, baked biscuits, and even cheese spread (Gonzalez et al., 2020).

Historical records suggest that mushrooms were cultivated in China as early as the seventh to tenth century, initially using very basic techniques such as placing steamed bran on a log and covering it with straw (Higgins et al., 2017). There is evidence that *wood ears* were cultivated in the seventh century, followed by *enokitake* in the tenth – eleventh century and *shiitake* in the eleventh–twelfth century (Singh & Mishra, 2008). There is also some evidence that *shiitake* mushrooms have been cultivated in Japan for 2000 years (Stamets & Chilton, 1983). The earliest record of mushroom cultivation in Europe goes back to the seventeenth century (Chang, 1977). Learning from farmers growing mushroom in rural France, the agronomist Olivier de Serres implanted mushroom mycelium into beds composed with materials in which the mushrooms selected for experiments were growing naturally. After the initial cropping, the same materials were reused to prepare new substrate for the following harvest. This was the first attempt to organise mushroom farming systematically, which led, as demand grew, to a large industry (Stamets & Chilton, 1983).

A critical step in building this industry was the development of a method to obtain mycelium from spores, through experimentation using the *button mushroom*, carried out by Constantin and Matroushot in the nineteenth century. The method was not disclosed to public until, in the USA, Fergusson replicated it and published his findings in 1902. Soon after, in 1905, another method of reproduction was discovered by Douggarou, when he found that mycelia could be grown from a tissue taken from a mushroom's cap. The public availability of both methods propelled the expansion of an industry specialised in the production of mycelium to be used in mushroom farming, allowing farmers to purchase – rather than grow inhouse - mycelia of specific mushrooms to start their cultivations (Agaricus, 2020). With mushroom farming establishing itself as a promising food industry, other discoveries soon followed. The process of using grain as a substrate to grow mycelium was developed in 1932, followed by the development of the pasteurisation of compost to eliminate organisms harmful to mushrooms as well as a procedure to accelerate the composting period (fast composting) by Sinden and Hauser, which is now commonly used (ibid). These processes as well as equipment for their development were progressively improved as the industry grew. Today, large mushroom farms can rely on computerised environmental monitoring systems and even automation in harvesting which allow an increase in production and cost reduction (Higgins et al., 2017). About three decades ago, computerised indoor climate control was introduced in the Netherlands and now is widely used in large farms (Sanchez, 2004). Currently, out of the known 2000 edible mushrooms, only 100 have been grown either experimentally or industrially, 30 mushrooms are commercially cultivated, and only 6 are utilised for industrial production (Chang, 2009).

4.3.2 Mushroom Farming: Functioning and Technologies

Techniques for the cultivation of mushrooms are illustrated in many manuals, some of which have been compiled particularly for small-scale farming in developing countries such as Zimbabwe (Kashungura et al., 2005), Namibia (Kadhila-Muandingi et al., 2008) and India (Singh & Mishra, 2008). The United Nations has also commissioned a manual for mushroom cultivation, which provides a complete compendium of the mushroom biology, nutritional value, and growing techniques available (Chang, 2009). In these manuals, mushroom farming is promoted as a viable source of income and food, particularly because of the availability at no cost of substrate in agricultural areas as well as the possibility of establishing a cultivation with limited resources and basic equipment.

The cultivation of edible mushrooms requires firstly the preparation of the mycelium, which will be used to generate the mushroom. This mycelium is grown in a substrate which, once colonised by the hyphae, is called spawn, a word derived from the old French verb *espandre* (i.e., expand). Subsequently, it involves preparing a substrate enabling the spawn to propagate (i.e., spawning) and finally triggering the growth of fruiting bodies (Chang, 2009). Higgins et al. (2017) divided mushroom farming into six components which occur over three phases of production. The six components include: Sterilisation; Inoculation; Substrate; Substrate Heat treatment; Climate Control; and Post-harvest practices. The three phases are summarised below.

4.3.2.1 Spawning

This is perhaps the most delicate phase, part of which is often carried out in specialised laboratories that provide spawn to mushroom farmers. It is the phase in which a culture of a mycelium is developed with no contaminants and therefore capable of developing into full mushrooms and fruiting bodies. This process requires sterilisation and accurate execution. A spore or a tissue from a specific mushroom must be selected and subsequently placed on a dish, in a medium such as Malt Extract Agar, which is a gelatinous agent, typically derived from seaweed. On the medium, a mycelium will form which needs to be transferred to the substrate where it will further grow. Industrial laboratories often develop spawn in a substrate such as cereal grains that are boiled and subsequently mixed with chalk and transferred to bottles for inoculation. Sterilised equipment and environment are key to successful development of the mycelium. Spores must be stored in a sterilised and appropriate space which is difficult to accommodate in low-resource mushroom farming.

Once the spawn is ready, it can be transferred to the substrate where it will finally grow and produce fruiting bodies. This substrate needs to be 'composted'. Different mushrooms will require different composts, varying from mixtures of agricultural waste to composted organic waste typically used for soil enhancement. In a compost with agricultural waste, this is shredded and processed in a mixer with water and other additives. Once ready, the compost is sterilised and inserted into containers. Alternatively, the substrate can be fermented with a process that is similar to organic waste composting, in which worms and microbes digest organic matter. Once ready, this compost too must be sterilised.

4.3.2.2 Substrate Sterilisation

The substrate must be sterilised to avoid larvae and other organisms damaging the mycelium. Typically, four methods can be used: sterilisation under high pressure, semi sterilization under low pressure, pasteurization by steam and pasteurization by hot water. Sterilisation under high and low pressure can require expensive equipment that small mushroom farmers may not be able to purchase. The substrate is subsequently transferred to different types of containers, depending on the mushroom and cultivation technique. In the projects visited, substrate was packaged in vertical plastic bags, into which the spawn was transferred. Vertical bags were hung on metal racks in dark grow rooms in which the temperature and humidity are controlled. Three to four weeks are necessary for the spawn to propagate within each bag. Subsequently each bag is punched in several points to allow the mushroom plant to fruit through these holes, a process triggered by modifying light, temperature, or humidity in the growing room. Compost can be packaged in different containers and laid on shelves, or even transferred to plastic buckets. These are all low-resource techniques that can be implemented on a tight budget.

4.3.2.3 Climate Control

Mushrooms grow within a small range of environmental conditions and the correct control of air humidity and temperature is fundamental to ensuring good yields. Environmental factors vary depending on the type of mushroom grown. However, an optimal temperature is in the range of 15–35 °C, with air humidity of 80–95% (Kadhila-Muandingi et al., 2008). Computer controlled monitoring of indoor environmental conditions is costly and typically deployed in large scale farming. Monitoring and air and temperature control systems are also very energy intensive although they can ensure high yields and, over the medium term, reduce operational costs. In the projects visited, low-cost humidifiers were used to achieve an optimal indoor environment, together with relatively basic ventilation systems.

Mushrooms harvested must be refrigerated in order to increase their shelf life and it is not uncommon to have high losses due to inappropriate conservation practices. It is however possible to dry mushrooms or to process and use them as an ingredient in a vast range of foods, including noodles, biscuits, or soup powder (Higgins et al., 2017). At an industrial level, mushroom farming is complex and requires high precision. Nevertheless, as the case studies demonstrate, it can be practiced with rather modest means, provided that precision and extreme care in sterilisation and environmental control is practiced.

4.3.3 Environmental Efficiency, Productivity and Market Context

Mushroom farming techniques are particularly appropriate for a circular economy-based system of food production. This is mainly because the substrate on which mushrooms grow is made out of agricultural waste or other food waste such as spent coffee grounds. Substrate based on spent coffee grounds is in fact particularly available in urban environments, where this is a by-product of the catering industry, available in large quantities and at short distances. A 2005 study (Tokimoto et al., 2005) estimated the global food production of spent coffee grounds to be 6 million tons. Spent coffee grounds are also used to generate energy. For example, Nestle' committed to reduce waste from its production by generating all the energy consumed in 20 factories by using spent coffee grounds (Campos-Vega et al., 2015). Agriculture and food processing industries produce other types of waste, which, if used as mushroom substrate, can improve their *circular* production. For example, olive oil pressing (Koutrotsios et al., 2018) and waste for wineries have been experimented (Murthy & Naidu, 2012) as viable options. As noted, once used, substrate can be composted or used as animal feed. In Asian countries mushroom farming is sometimes integrated with rice cultivation, with rice straw used as a substrate (Rosmiza et al., 2016). Another advantage in the utilisation of agricultural waste for mushroom farming is that, if unused, this is typically burnt and increases air pollution (Chang, 2009). In 2009, 0.25 billion tons of straw were burned in China alone (Grimm & Wösten, 2018). Mushroom cultivation is therefore particularly appropriate for developing countries with big agriculture outputs, such as African and Indian countries, where these raw materials are available in large quantities and are cheap (Chang, 2009).

In farms equipped with environmental control systems, energy is one of the main contributors to GHG emissions. An LCA study on a French mushroom farm (Dorr et al., 2020) found that energy was the most impactful category, followed by transportation, considered through the whole life cycle of the farming process. As the study mentions, other LCA studies on mushroom farming reached similar conclusions, finding energy for climate control and machinery as the most impactful factor (see Leiva et al., 2015; Robinson et al., 2018). However, another study found that pre-farm activities such as delivery of materials (e.g., substrate) can generate the highest impact (Gunady et al., 2012). Efficiency in production is generally quantified as yield per substrate weight. Typically, 1 kg of mushroom requires 5 kg of substrate (Rosmiza et al., 2016). But substrate can be reused for more than one harvest. Although the reuse may result in lower yields, a study by Cunha Zied et al. (2020) reviewed several techniques to ensure high productivity with spent compost. Productivity, growing techniques, location of suppliers, and distribution will inevitably play a significant role in the environmental efficiency of the mushroom farms. In fact, a conclusion from the study by Dorr et al. (2020) was that higher yield through improved sanitation practices, resulting in minor losses of mushrooms, would reduce the overall impact of the farm. Mushrooms, just as all organisms, can be damaged by several pests. For an organic cultivation, the best way to manage pest control is pasteurisation and sanitation in all phases of production.

Water use for mushroom cultivation is difficult to determine because it will vary depending on the growing technique, the equipment, and the mushroom species. A study from Sure Harvest (2017), a Californian software company for agricultural production, suggests that 1.8–2 gallons of water is required to produce 1 pound of button mushrooms. This corresponds to 15–16.7 L per 1 Kg of mushrooms. However, different composting techniques may have different water requirements. In the LCA study by Dorr et al. (2020), water consumption for 1 kg of Oyster mushrooms, both for composting (with spent coffee grounds) and cultivation was 26.36 L, considerably higher than the previous evaluation. Finally, in a publication on mushroom farming by Rotterzwam, an urban mushroom farm in Rotterdam which is documented as a case study in this book (Rotterzwam, n.d.), it is suggested that 15 L of water are necessary to grow 1 kg of Oyster mushrooms. The publication does not provide a breakdown of the water used for composting and cultivation.

The cycle of production of mushrooms is fast, taking 5 or more weeks, depending on the mushroom. Annual production by weight can therefore be substantial even for small farms in developing countries, with basic equipment and low investment. A small mushroom farm requires minimal land and input. A study by Higgins et al. (2017) suggests that in order to grow 80–130 kg, a Cambodian farmer would spend USD 55 – USD 70, with an average net profit from mushroom sales (quantified at USD 2.87 per kg) of up to USD 200. Likewise, a study of small mushroom farmers in Bangladesh suggests that with investments ranging between USD 65 and USD 1,285, it is possible to generate monthly returns ranging from USD 30 to USD 256 (Easin et al., 2017). FAO is promoting mushroom farming in developing countries such as Uganda (FAO, 2017) and Thailand, a country in which a pilot mushroom farm for disabled farmers was tested, which resulted in a manual compiled specifically for this audience (Hanko, 2001). As we will see through the case studies, conditions in Europe are radically different and, even with good yields, because of high production costs, returns can be generated only through a diversification of activities linked to production, including consultancies and tutorials on mushroom farming techniques.

At an industrial level, the mushroom industry is booming. In 2015, the global demand for mushrooms was estimated to exceed USD 35.03 billion and projected to reach USD 59.48 billion by 2021 (Higgins et al., 2017). China, a country with the greatest output worldwide, produced more than 8.7 million tons in 2002, accounting for 71.5% of global output (Yoo et al., 2016). This is the fifth largest agricultural sector in the country, employing 25 million farmers, with a value (as of 2011) of USD 24 billion, and technological and bio-technological developments enabling higher efficiency and lower costs (Zhang et al., 2014). Other countries leading in this production are the USA and EU, but medium income countries are quickly developing this industry (Higgins et al., 2017).

Chapter 5
Methodology of the Study: How Success Is Measured

Abstract This chapter outlines the methodology followed for the investigation presented in the book. It is a mixed methodology although predominantly qualitative. A literature review plays a big part: grey and peer-reviewed literature was used to gather information about the history and state-of-the-art of soil-less methods. Other scientific literature contributed to develop an understanding of the complex relationship between technology and society, which in turn is key to understand the relationship between the urban farmers of the case studies and soil-less technologies. A web-search was instrumental to identify the projects visited and analysed as well as to identify the scale of interest of the public for soil-less technologies. Case study analysis was used to generate findings from specific projects. Quantitative and qualitative data were collected through semistructured interviews, and organised according to a basic grid of analysis. This comprises four areas: (a) motivations of urban farmers; (b) outcomes of the project; (c) productivity: and (d) environmental performance.

5.1 General Methodology

This chapter outlines the methodology for this investigation, before proceeding to present the case studies and discuss their material and immaterial productivity. The methodology used is predominantly qualitative and based on several methods. Literature review plays a big part; it was used to develop the conceptual framework within which the topic of the book is situated. Existing studies on urban agriculture in its several manifestations and models, mainly over the last two decades, were essential to an analysis of its evolution and the identification of present and future trends. Grey and peer-reviewed literature was used to gather information about the history and state-of-the-art of soil-less methods. Other scientific literature contributed to develop an understanding of the complex relationship between technology and society, which is an important one to consider in order to frame the relationship between the urban farmers of these case studies and soil-less technologies. A web-search was instrumental to identify the projects visited and analysed, and to gather insights about the online soil-less farmer community (see Chap. 7) as well as the

S. Caputo, *Small Scale Soil-less Urban Agriculture in Europe*, Urban Agriculture, https://doi.org/10.1007/978-3-030-99962-9_5

extent to which urban farmers are modifying soil-less technologies, thus producing grassroots innovation. Case study analysis was used to generate findings from specific projects. The case study research method enables an in-depth investigation of 'few units with multiple variables' (Krusenvik, 2016), the results of which cannot be generalised because of their discrete and context-dependent nature, but they can be used to validate a hypothesis (Flyvbjerg, 2006). Case studies are in fact used partly to discuss hypotheses generated from the literature review in terms of objectives and purpose of urban agriculture. At the same time, multiple case studies are used to gather qualitative data and some quantitative data, based on which some directions and characteristics of this trend in urban agriculture are discussed in Chap. 8. This does not lead to generalised findings, but rather to the identification of potential benefits and drawbacks. Data collected through semi-structured interviews are organised according to a simple grid of analysis, which is composed of the main questions asked to the interviewees and/or of the factors that the observation carried out during visits attempted to ascertain. This grid is based both on literature review and on the stated scope of the book, as explained below.

The reason to develop a qualitative analysis is that the case studies selected for this book share a strong social agenda that drives their activities. Measuring their success by using indicators strictly connected to quantifiable food production and environmental efficiency would fail to capture the actual benefits that these case studies strive to achieve; outcomes of a quantitative analysis alone can point to ineffectiveness in terms of food harvested and inputs required for food production, misrepresenting the real value of these projects. In fact, for some of these case studies, higher productivity or environmental efficiency is not a priority or the strongest motivation for adopting soil-less technologies. Another very good reason is that, in some case studies, a record of the food harvested, and the inputs utilised was not available, which, again, suggests that such records are not regarded as important when compared to the level of community engagement that the soil-less projects can generate. Some of the farmers managing the projects consider the soil-less units as demonstrators, just one of the possible techniques for growing food in cities that can be included in their gardens and farms to inform volunteers and visitors. They can also be used to engage young students, who are likely to be attracted by unconventional technologies of food production when compared to conventional horticulture. Hence, rather than measure efficiency, the focus is on effectiveness in attaining the goals that the community group or small enterprise in each case study has set. The evaluation is based on anecdotal evidence and observation, therefore mainly qualitative and with a degree of imprecision. Although mainly discursive, this evaluation serves the purpose of identifying the motivations behind the choice of soil-less technologies and the impact that these have on practicing urban agriculture as well as the broader impact generated on the volunteers and/or visitors. One of the aims of this book is to trace the new directions that urban agriculture is taking and, in doing so, anticipate new possible scenarios. The approach for this exercise is to identify elements that are shared between the case studies and validated in the literature to subsequently attempt to identify the consequences over the medium-term, should these elements become consolidated and more powerful (see Chap. 9).

Case studies were selected after an online search and consultation with experts, developed in the book's initial stage of drafting and with a non-systematic approach. To our knowledge, there is no open access register or list of soil-less enterprises – let alone community-led urban agriculture projects - that can be found through national specialised groups or associations. Google was the search engine used, with keywords or strings of keywords that included the following in different combinations: soil-less, hydroponic, aquaponic, mushroom farm and farming, community gardens, urban farms, urban agriculture, urban farming, and the name of European countries (e.g. France, Spain, Italy, etc.). An additional source of information was provided through consultation via email, with academic and practice-based contacts as well as international research groups in this subject area, with pages on Research Gate, Facebook and LinkedIn. Some projects were also identified through research projects and the networks connected to them, such as the one generated through the COST Action 'Aquaponics'. An important factor of this search, which reflects the nature of this book, is the attempt to document case studies that are not so well known and that have not been documented in the relevant literature. Being different, these case studies demonstrate the extent to which the soil-less urban agriculture is diversifying, and the possible directions that it may take.

The initial list of projects (included in Appendix A) was subsequently filtered in order to ensure that the case studies selected were in line with the scope of the book. Hence, projects shortlisted were either community-based or small enterprises with an explicit focus on sustainability. The latter is, however, an ambiguous category. A small enterprise, in the definition of the European Commission, is an enterprise that has below 10 staff members and a turnover of below 2 million euro (European Commission, 2020a, b). Social enterprises are enterprises that are socially driven and reinvest profits to achieve social objectives (European Commission, 2020b). The commercial enterprises selected for this study are smaller than the description of small enterprises provided by the European Commission. This is partly because it is not easy to define the thresholds for each scale. Villerroel et al. (2016) suggest that a large aquaponic farm is above 1500 m^2. They also distinguish between producers that harvest an annual quantity of fish of less than 100 kg and more than 1000 kg, perhaps offering a way to categorise small and large farms. This leaves a big gap for medium farms. One farmer from a case study expressed the opinion that economic viability for medium soil-less farms is not attainable. High investments for infrastructure and specialised labour required for indoors farming make it difficult to break even without an economy of scale. Conversely, small urban farms with 2–3 farmers require smaller capital and management costs; at that scale, yields and income are sufficient to cover costs.

For the purpose of this study, three categories were considered: (a) soil-less projects (community-led projects that are not selling their food and/or are limited in size (below 50 m^2) and scope of production); (b) small farms (for-profit enterprises that employ no more than three people and are below 750–1000 m^2 of production area); and (c) medium farms (for-profit enterprises that employ more than three people and are below 1500 m^2 of production area). Some case studies do not completely fit into the categories as defined. One of the community-led projects (Huerto

Lazo – see Sect. 6.2) is larger than 100 m^2 and sells part of the fish farmed. However, the agenda of this project is evidently non-commercial, and its organisation is far from being entrepreneurial. Another case study, the aquaponic medium farm GrowUp (see Sect. 6.9.1, lies beyond the focus of this study (which is on small scale urban agriculture). It is included since it offers a term of comparison for food productivity within the case study sample (see Chap. 8) and because of its exceptional social agenda which shows how commercial production may not be detrimental to social sustainability within an enterprise's vision. In addition, a few educational projects were included (i.e., projects whose main objective is to use soil-less systems as demonstrators rather than for production only). Therefore, the definition of each category must be understood as rather loose by necessity, since the nature of each of the case studies is extremely varied and complex in ambitions and operations. Appendix A includes projects and farms that qualify under these categories. It also includes experimental projects and R&D facilities, and technology providers. As mentioned, the final list of projects is not the result of a systematic search, is not exhaustive, and yet is representative of the rapidly evolving scene of soil-less farms and projects which has changed and grown during the drafting of this book. The list is therefore only hinting at the breadth of and potential for this scene to grow and diversify in several possible directions.

Projects and small farms identified in the list were contacted but only a few responded and agreed to be visited. The resulting final sample of case studies comprises 1 hydroponic small farm, 2 mushroom small farms, 5 aquaponic community-led projects, 3 aquaponic small farms (1 of which is which is currently functioning as a R&D facility), 1 aquaponic medium farm and 2 educational hydroponic projects (see Table 8.1, Chap. 8). The initial intention to have a similar number of hydroponic and aquaponic projects and small farms, and a few mushroom small farms had to be reconsidered because, as explained above, not all the projects agreed to be interviewed. Yet, although most case studies use aquaponics technology, we believe that the sample succeeds in offering a picture of small scale soil-less urban agriculture as a whole. In aquaponic farms, the hydroponic unit is as important as the aquaculture one, and considerations of farmers on the former are amply documented in each case study.

An even geographic distribution of the projects was sought. The search mentioned above showed a higher concentration of projects in Northern Europe, with a smaller number in Southern and Eastern Europe. As a result, projects identified are predominantly in North European countries (UK, France, Germany, Belgium, the Netherlands and Sweden) with only a few in Spain and Italy. It must be stressed that many of the projects included (and sometimes small scale farms too) are small and hardly 'visible' on the web. Some of them, for example, do not have a dedicated website and, even when they have one, this is not easily identifiable with a Google search. Difficulty in retrieving information on soil-less projects makes it hard to estimate the scale of this phenomenon, which may be larger than it appears. The final selection of case studies was visited between 2018 and 2019. Two case studies were visited 'remotely' in 2020 via video call, enabling a virtual visit of the soil-less units through video camera.

5.2 Scope of Evaluation

There are several evaluative studies focusing on the productivity of urban agriculture, which were developed over the last two decades. They attempt to demonstrate urban agriculture's contribution to food security and its significance for the attainment of urban sustainability. Food growing is part of a tradition of urban gardening that in the post-war period has been attractive for those who enjoyed gardening as a hobby and for its social benefits (Acton, 2011). At the turn of the millennium, with urban sustainability becoming ever-more important, there is the need to prove that urban agriculture matters also for the contribution it can generate for food security. In this vein, a study published in 1999 by Garnett, suggests that London can produce food to meet 18% of Londoners' vegetable intake. If this seems a small share compared to the whole city's demand, other studies reached different and more optimistic conclusions. Amongst the most recent ones, Ackerman et al. (2014) estimated that vacant land in New York (about 49,884 acres) is not enough to make the city self-sufficient. However, when the extended metropolitan area is considered, there is sufficient land to feed between 58 and 89% of the city's population. More optimistically, Saha and Eckelman (2017) evaluate that, with plots at ground level and rooftops, 17.4% of the urban area is available in Boston, which can produce enough to feed the entire city with a sufficient supply of fruit and vegetables. Similar to the shift outlined in Chap. 2, in which the scope of urban agriculture broadens to become a widespread and also commercial practice, we can see in these studies a shift of spatial focus from green land to rooftops to indoor cultivation. A case in point is the study by Nadal et al. (2017), according to which, with all suitable rooftops equipped with greenhouses, the city of Rubi, in Spain, can produce 50% of the local demand for tomatoes. Vertical farming is an intensive hydroponic farming type, particularly suitable for cities because of its significant land-efficiency, with some farms already in operation in the US, the Netherlands, Japan, Singapore and South Korea (Kalantari et al., 2018). Al-Chalabi (2015) suggests that this can be a type of farming particularly suitable in climates and urban spaces with high solar radiations.

A shift in the perception and the scope of urban agriculture is also reflected in recent attempts to measure its efficiency. For example, Farming Concrete uses citizen science to gather data on types of crops planted and harvested; waste management and the quantity of compost produced; numbers of volunteers; the time worked and the number of attendees at events; perceived improvements in mental and physical health; and economic data on produce sales and food donations (Gittleman et al., 2012). The objective is to measure the wider range of benefits from urban agriculture by collecting quantitative and qualitative data, recognising the socio-political dimensions of the practice. The study that this book documents aligns with this idea of urban agriculture. In order to develop the case studies, semi-structured interviews with urban farmers were conducted, with informal conversations aimed at identifying the organisational structure of each project and its history, and at understanding the approach guiding each project in four main areas: (a) *motivations* behind the inclusion of soil-less technologies in the community project; (b) *outcomes*: material and social benefits generated (education, health, community-building and economic – the latter, when relevant); (c) *productivity*; and (d) the *environmental performance* of the soil-less units and the environmental awareness of the farmers.

The first area tries to cast some light on the motivations behind a choice that is not in line with the conventional way of practicing horticulture in cities. Although, as this book maintains, urban agriculture is changing fast, most urban gardeners and farmers enjoy working outdoors, with their 'hands stuck in the soil' (Caputo et al., 2020). Many of them embrace environmental causes (Scheromm, 2015), therefore using their gardens as places that can have a positive and real impact on the environment. They grow plants that attract pollinators, they harvest rainwater, they increase biodiversity, and they are aware that, by expanding the urban green infrastructure or by making sure that part of this infrastructure is preserved, they contribute to improved local and global climatic conditions. Interviews aimed at understanding how the interviewees reconcile these objectives with the choice of producing food through soil-less technologies and indoors. Conversations were not designed to enquire specifically about this apparent contradiction; they were intentionally generic in order to avoid bias in responses and understand whether the choice was perceived as possibly clashing with environmental objectives. In most of the case studies, there was little awareness about this issue.

The second area shifts the focus from the perceived motivations driving choices to the actual outcomes that such choices generate. These are connected to the agenda of the project: enterprises (both for-profit and social) generally aim for profitability; community projects can use the soil-less units as, for example, demonstrators for educational purposes. Increasingly, community gardens and city farms put health and wellbeing (physical and mental) at the core of their activities, with some specifically shaping their agenda around mental wellbeing (Malberg Dyg et al., 2020; Baur, 2020). In the UK, the association Care Farming (www.carefarminguk.org) supports farms that use their farming practices for therapeutic purposes. The organisational model of community farms and gardens is based on volunteers and trainees, who exchange time and labour, often having in return health benefits (physical exercise and mental wellbeing) as well as an opportunity to socialise. These places have a rich agenda of community engagement with workshops and events to increase their outreach. Regardless of the specific aim of each project, interviews intended to identify in what way practicing urban agriculture with soil-less technologies contributed to or deterred achieving the stated aims and if there were any other unexpected benefits.

The third area focuses on productivity of soil-less systems. This is a difficult indicator to investigate due to the lack of data from case studies as well as benchmarks against which their productivity can be assessed. A further level of difficulty is represented by the diversity of crops that can be grown, each one with a possible distinct benchmark of soil-less productivity, very few of which are available. The different objectives between case studies is also problematic; commercial farms are highly motivated to maximise production and attain financial viability; community projects are not. It would be incorrect to compare the two types. The few data on productivity cannot be interpreted literally but must be evaluated against the characteristics of each case. Commercial farms are more likely to keep a record of the inputs and harvests, which can help identify areas for improvement. Some community-led projects provided data too although some are estimates rather than data reliably gathered. Despite these difficulties, using data gathered from the case studies that made them available, productivity is benchmarked against that which has been reported in Chap. 4 (see Tables 4.1, 4.2, 4.3, 4.4 and 4.5). These are data indicating the productivity of industrial and experimental units which are not congruent with the cases here examined. Yet, the comparison points to some interesting findings. Moreover, other benchmarks are also considered that are taken from FAO reports on small scale soil-less projects.

The fourth area aims at identifying the degree of environmental awareness of the urban farmers and the effective environmental performance of the soil-less units. These units require a different range of inputs if compared to in-soil food production. They can use much less water and at the same time consume much more energy. LCA studies have suggested that hydroponic and aquaponic units can generate a higher Global Warming Potential (GWP) than conventional agriculture, singling out energy as the factor contributing to this impact. The term 'environmental awareness' captures knowledge, understanding and behaviour of gardeners and farmers in practicing horticulture, resulting in a low or high level of inputs. Measuring environmental awareness requires the availability of data as well as benchmarks. Similar to productivity, both have proved to be problematic. As noted, some of the projects included in this book do not have data available. Community-led projects do not generally seek scientific validation for their sustainability. Their organisational structure (with a high turn-over of volunteers) makes it difficult to rely on people who can record inputs and outputs on a regular basis, and for a sufficiently long period. Their social agendas do not prioritise maximisation of production and, because of the strong focus on the social benefits generated through gardening, these may not be particularly attentive to environmental efficiency of such a production. The evaluation of the environmental efficiency is therefore generally not quantitative, but rather qualitative and based on responses. It is also based on observation and the real characteristics of the project. All urban farmers and gardeners consulted are driven by sustainability principles, but this intention does not always translate into effective environmental strategies. Case studies are therefore evaluated for their environmentally positive or negative characteristics. A few case studies have provided data, enabling an evaluation of their performance. Some benchmarks have been identified in the literature, referring to the efficiency of experimental units or industrial soil-less farms, which may not be appropriate for comparison with small scale soil-less projects and farms. Yet, in Chap. 8, for those case studies that provided sufficient information, a brief comparative analysis is carried out. Factors indicating the environmental efficiency of each soil-less project include water consumption, energy consumption, onsite energy production, synthetic nutrient added, and fish feed quantity.

In Chaps. 6 and 8, case studies and discussion are organised in sections that follow the four areas mentioned above (motivations; outcomes; productivity; and environmental performance). Due to the relatively small number of case studies covering the three different soil-less technologies as well as the small dataset available, the quantitative analysis provided in Chap. 8 focuses on hydroponics and aquaponics, together. This is because only one hydroponics case study could be found and visited. Mushroom farming is also considered, although to a smaller extent. The qualitative analysis developed on the basis of the observation and interviews is richer than the quantitative one. Both analyses offer an interpretation of the phenomenon, a direction of evolution of community-led soil-less urban agriculture, and how this can contribute to the attainment of some of the key objectives of this practice generally.

Chapter 6
Case Studies

Abstract The case studies presented in this chapter were visited between 2018 and 2020. Community-led projects and enterprises are equally represented in the case study sample, which includes 5 community-led projects and 5 small or medium enterprises, based on aquaponic or hydroponic technologies. 2 of these projects were closed during the drafting of the book. However, they were included in order to document the vulnerability of small-scale initiatives, which must struggle with economic viability and organisational issues. The case studies also include 2 mushroom farms and 2 organisations that promote hydroponics as an educational tool and provide hydroponic training courses. The latter were selected to document the inclusion of aquaponics in school curricula, which has been implemented in several institutions. This range of diverse projects well represent the breadth of activities and the interest that soil-less technologies attract. Each case study gives an overview on motivations that led farmers to start the project as well as a description of the outcomes of the project. Data on productivity and environmental performance are also included for those case studies that were able to provide them. The stories of the soil-less urban farmers and their achievements constitute an important and valuable account of the way urban agriculture practice, attitudes and values are responding and adapting to societal changes.

The case studies presented in this chapter were visited between 2018 and 2020. As mentioned in Chap. 5, the selection of the case studies was dictated by the willingness of the community-led project and farm managers to be interviewed, therefore opportunistic rather than planned. Community-led projects and enterprises are equally represented in the case study sample, but only one hydroponic farm is included against nine aquaponic projects and farms. The predominance of aquaponics in the book does not necessarily reflect the actual European context. As summarised in the table in Appendix A, other small hydroponic projects were found in the search and selection of suitable case studies. Also, as Chap. 7 suggests, on the web, there seems to be more interest in hydroponics than aquaponics. However,

S. Caputo, *Small Scale Soil-less Urban Agriculture in Europe*, Urban Agriculture, https://doi.org/10.1007/978-3-030-99962-9_6

although hydroponic-only projects and farms are under-represented in this chapter, the discussion on both technologies is equally well developed. Ultimately, hydroponic technology is at the core of aquaponic community-led projects and enterprises. Issues such as crop selection, productivity and environmental efficiency in hydroponics must be dealt with in aquaponics too. Likewise, both technologies have in common issues of safety, quality and customer's acceptance of food grown in controlled environments. No community-led mushroom project was found in the search, but the two mushrooms farms are small and sustainability-driven; the discussion on the advantages and drawbacks of adopting soil-less technologies in an urban environment is well developed for this specific technology too. The case studies include one community-led project and one farm that were shut down before ending the draft of the book (Sect. 6.9). Two organisations that promote hydroponics as an educational tool (Sect. 6.12.1) and provide hydroponic training courses (Sect. 6.12.2) are also included. The aforementioned community-led project and (subsequently closed) farm were included to document the vulnerability of small-scale initiatives, which must struggle with economic viability and organisational issues. Many soil-less farms had to face closure shortly after starting up. A case in point is Urban Farmers in Den Haag, as mentioned in Chap. 1. These two case studies were active when interviewed but their closure offered the opportunity to elaborate on this important issue. The latter two were selected to document the inclusion of aquaponics in school curricula, which has been implemented in several institutions (Junge et al., 2019). Both are instrumental for representing the breadth of activities and the interest that soil-less technologies attract. As noted, not all case studies were able to provide quantitative data. Yet the stories of the soil-less urban farmers and their motivations in embracing this technology constitute an important and valuable account of the way practice, attitudes and values are responding and adapting to societal change.

Community-led soil-less projects.

6.1 El Milagro de los Peces (Community-Led Aquaponic Project)

Director: Pepe Lobillo
Location: Instituto de educacion secundaria Joaquin Romero Murube, Poligono Sur, Sevilla, Spain
Website: http://www.iesromeromurube.es/
Visited in: May 2019

Motivations Poligono Sur is one of the poorest neighbourhoods at the outskirt of Sevilla, a city in south Spain with a population of about 700,000 inhabitants. The neighbourhood is home to 55,000 people, 40% of whom are jobless and an estimated 80% work in the informal labour market. Poligono Sur needs maintenance, its streets are littered, there is a lack of public space and green areas, and a general atmosphere of dereliction and lack of safety. The secondary school in this neighbourhood, Instituto Joaquim Romero Murube, has many students with a difficult background, and the school has developed special strategies to attract and retain such students. In particular, the school offers vocational activities for those who have little interest in the academic curriculum. They have a school garden where food is grown and where plots are made available to be rented by the students' families. In the school's garden, an aquaponics facility has been installed, which, at the time of the visit, had been working for 3 years. It is directed by Pepe Lobillo, who, on behalf of the Andalusian Local Government, is a social worker at Poligono Sur.

The story leading to the opening of this aquaponic facility is an interesting one and starts many years before this project was initiated. Pepe Lobillo is a social educator and veterinary who holds a PhD in aquaculture. He has always been involved in social programmes supporting the disadvantaged communities in Sevilla, and his collaboration with local authorities led him in 2012 to fund the initiative Verdes del Sur, which aimed at finding and making use of derelict land in Poligono Sur to start new community gardens and allotment sites. Over the initial 2 years, more than 300 families (2000 people) joined this initiative. Eventually, Verdes del Sur found suitable places for growing, but planning consent from the municipality for their occupation as food gardens was slow to come. The coordination of the group was also difficult as families often disagreed on strategic directions to take. As part of Verdes del Sur's engagement activities, a group of members led by Lobillo visited the facilities of the Escuela Técnica Superior de Ingeniería Agronomica at the Universidad de Sevilla, which also includes an aquaponic unit, used as a training facility. To date, only a few Schools of Agronomy in the country can offer an aquaponic training facility. In Sevilla, Professor Victor Fernández-Cabanás is championing aquaponics and is advising and supporting Pepe Lobillo in his efforts.

One of the families visiting this facility was particularly impressed by the aquaponic technology and decided to build a small-scale unit for a single household, in their backyard. They asked Pepe Lobillo to help and received support from him and from Professor Victor Fernández-Cabanás. Lobillo had previous experience of small-scale aquaponics, developed during a trip to Mexico, in which he had the opportunity to observe chinampas built by local people. He was also fascinated by the aquaponics networks established through the Australian Backyard Aquaponic website,[1] where resources and guidance can be found for self-build domestic units. This project was in line with one of the objectives of Verdes del Sur which was the attainment of food security and sovereignty, in a poor neighbourhood of Sevilla where this was much needed. The result was a pilot project run by this family and named Milagro de Los Peces (Miracle of the Fish). The project is documented in some videos on YouTube,[2] which present a 4 m^2 unit comprising three fish tanks (about 1000 L each) with ebb and flow trays on top (Verdes del Sur, 2020), constructed following the FAO manual of aquaponic unit for single households (Somerville et al., 2014).

[1] www.backyardaquaponics.com

[2] https://www.youtube.com/watch?v=I47zy3LgzeU

The project came to a halt when the member of the family who primarily managed the unit had to abandon the household due to legal issues. Using the experience developed in this pilot project, Pepe Lobillo proposed to the members of Verdes del Sur to build a small unit on the grounds of the local secondary school. On those grounds, the association had already helped organise successfully the community garden and the allotment plots; the aquaponic unit was therefore a welcome addition. Verdes del Sur launched the project on Goteo – a crowdfunding website - and succeeded in collecting €10,000, which paid for materials, labour and equipment.[3]

Outcomes Although one of the main motivations behind Milagro de los Peces was food security, its final outcome is predominantly educational. The aquaponic unit at the Instituto Joaquim Romero Murube acts at the same time as a demonstrator for local households, showing that a domestic aquaponic unit can be self-built, and as a training facility for the students of the school. As, initially, students did not show particular interest in engaging with aquaponics, a group was selected by the school's teachers which included students who were underperforming academically, lacking motivation, interest and focus. They were asked to join a team working 1 hour per day in the aquaponic unit. Not all those selected developed an interest in this activity and some left. The persistence of the school in identifying new potentially suitable students was eventually successful, and a team of ten students is now formed. From the school's perspective, the overall objective is not only to equip students with life skills that may be useful in the future, but also to develop a dynamic school ethos and a pedagogical model in which activities help students pursue vocational options, when appropriate. The focus of such options on urban food production seems particularly relevant in this neighbourhood.

A model was built with some students (Fig. 6.1), which helped them understand the components of the aquaponic unit, and their use and functioning. The greenhouse that was subsequently built contains two 1000 L IBC (Intermediate Bulk Containers in plastic and metal) used as fish tanks, each one connected to a hydroponic unit approximately 3 m long. The hydroponic unit consists of a double PVC tube with regularly spaced holes (approximately every 25 cm) in which rockwool blocks are inserted. In between the tubes and the tank, other plants are grown in a container filled with expanded clay pebbles (Leca). The aquaponic system includes a bio-filter, which also acts as a grow-bed, and a submersible pump of 32 watts with a maximum flow rate of 5000 litres/hour. The greenhouse is 9 × 5 m and 3.5 m high at the highest point (Figs. 6.2a and 6.2b) with a frame in steel tubes, and an envelope of PET membrane. Between the membrane and the steel structure, a steel grid prevents vandalism. Since the greenhouse was completed in 2016, there has been an attempt to break into it. Not much damage was done on that occasion but, during the repair work, the grid was added. At the time of the visit, only one of the two fish tanks and the hydroponic unit were functioning. In 3 weeks, the second unit had to be emptied for the summer, when the school closes for the holiday and students

[3] https://en.goteo.org/project/el-milagro-de-los-peces

Fig. 6.1 Model of the aquaponic unit

Fig. 6.2 (a, b) The aquaponic greenhouse in the grounds of the school and its interior

can no longer manage the aquaponic unit. Furthermore, there is no natural ventilation in the greenhouse and, in hot months, the indoor temperature peaks to 50 °C or above, making the greenhouse unsuitable for fish and plants. For this reason, between June and October, the roof membrane is temporarily substituted with a 40% shade mesh, which keeps the indoor temperature below 40 °C. During the winter months, the IBCs are heated through solar panels, ensuring that the water temperature never drops below 13–14 °C. When the outdoor temperature approaches 0 °C, a submersible heater is activated at night for approximately 10 hours. This, however, is a rare event: in Seville, the outdoor temperature rarely drops below 5 °C.

Lobillo was particularly active in engaging students in side projects, including a small 'green wall' and another small aquaponic unit. Videos documenting and promoting the aquaponic unit and these other projects project were edited and uploaded on YouTube.[4] The project is quite well known in the neighbourhood and has had some media coverage, thus acting as a tool for dissemination.

Productivity The small unit demonstrated good productivity. This was also due to the prior experience of Lobillo and his collaboration with the Universidad de Sevilla. This collaboration between practitioners and academia was useful not only for sharing knowledge and technical expertise but also for developing and implementing a reliable system of data collection that could demonstrate the potential of self-build aquaponics to meet household needs and be resource efficient. Although this potential is documented in the FAO book on aquaponics (Somerville et al., 2014), data collection within this project can confirm FAO's data and, more importantly, assess productivity in a developed country context, in which motivations and socio-political conditions are radically different. To this end, data on production and resource consumption of the unit were published in a peer-reviewed journal (Lobillo-Eguíbar et al., 2020). Over the academic year 2018–2019, 177.66 kg were harvested, consisting of 22 types of crops. Crops included a mix of fruits (i.e., watermelon, strawberry, melon, aubergine, courgette, cucumber and two varieties of tomato), vegetables (i.e., three types of lettuce, chard, onion, three types of pepper, broccoli, cauliflower, cabbage, potato, Chinese cabbage and pumpkin) and herbs (stevia and basil). The variety of crops is considerable and includes some that are not typically grown with hydroponic technology such as potato. Crops were selected with the objective of demonstrating that the small scale unit can meet the dietary requirements of a family, with a diversity of produce enabling a nutritionally healthy diet. The fish selected for this project was a red hybrid tilapia and the total harvest over the same period was 33.5 kg. The food was distributed between families and students who helped in the aquaponic project, some teachers and the two researchers (Lobillo and Fernández-Cabanás). No food is wasted and when the tanks must be emptied before the aquaponic unit closes for the summer months, the fish is frozen and stored, and distributed when the occasion arises. A microbiological analysis of plants and fish was performed. All tests were negative for Salmonellas spp., Escherichia coli and Listeria monocytogenes.

Environmental Performance The construction materials utilised for the greenhouse are basic and do not protect from outdoor temperatures, resulting, during the coldest months, in reduced production of fish and vegetables. Clearly, performance of the unit could improve with higher investment allowing, for example, a more efficient greenhouse envelope. However, this low-tech aquaponic prototype is effective in demon-

[4] https://www.youtube.com/watch?v=TFdV_jfyF6w&feature=youtu.be; https://youtu.be/MFirY8biYhc

strating the possibility of low-budget implementation of this technology, which is key to its uptake within the low-income communities living in Poligono Sur.

The aquaponic facility had a recirculating water volume of 1800 L, with a daily water consumption of 27.15 L per day and 7.2 L per day of water discharged. 15% of the total volume came from rainwater. A pump, two air compressors and, in winter, a thermostat and a heater were electrically operated with a total consumption of 334.5 kWh per year. The heater was operated through a solar panel which was completely self-assembled.

Of the 16 nutrients that plants need to grow, there were deficiencies in 3 nutrients in the water of the aquaponic system: iron, potassium and manganese. Iron was added to water in the form of chelated iron, authorised in European agricultural regulations. Potassium and manganese were added by foliar application of potassium sulphate and manganese sulphate solutions. A summary of productivity and performance can be found in the Table 6.1 below.

Table 6.1 Milagro de los Peces: productivity and environmental performance over 1 year (2018–2019)

Milagro de los Peces	Community-led aquaponic project
Objective	Educational; self-supply
Employees/ volunteers	1 organiser + about 10 volunteers
Total size of the greenhouse	45 m^2
Production units and area	Fish tank: 1800 L Hydroponic unit: 4.56 m^2 (grow bed, NFT unit and DWC unit)
Initial estimated investment	€4000 for the greenhouse. The cost of a similar backyard aquaponic unit could be € 1342
Energy consumption	334.5 kWh/year
Water added daily	27.15 L/day
Density fish	Minimum density – 1.6 kg/m^3 Maximum density – 20 kg/m^3
Density plants	20 plants per m^2
Fish harvest	33.5 kg/year
Fish variety	Tilapia
Crop harvest	177.66 kg/year
Crop variety	Watermelon, strawberry, melon, lettuce (3 types), chard, tomato (2 types), aubergine, cucumber, onion, pepper (3 types), broccoli, cauliflower, cabbage, potato, courgette, Chinese cabbage, pumpkin, stevia and basil.
Fish feed / cultivated area	Annual average of 45.5 ± 26.8 g fish feed/m^2 plant culture area

6.2 Asociación Huerto Lazo (Community-Led Aquaponic Project)

Directors: Ulrich Eich; Amancio Jimenez
Location: Cajiz, Málaga – Spain
Website: www.huertolazo.eu
Visited in: May 2019

Motivations This project started from a personal interest of the director, Ulrich Eich, in fish farming, his concern about the scarcity of resources worldwide and the particular social activity at the core of the German association Huerto Lazo. Ulrich helped setting up this association, which has the stated objective of supporting young adults with learning difficulties to integrate into society. Ulrich lives in a house located in the peri-urban, agricultural area on the outskirts of Cajiz, a town located 27 km from Malaga. Young adults following the Huerto Lazo programme are offered to spend a period of time at his house, with the opportunity to work in the aquaponic unit, benefitting from manual labour, proximity to nature and acquiring new skills. The house and sheds next to it – in which the aquaponic unit is located - are surrounded by 5 hectares of exotic fruit trees. Amancio Jimenez, the owner of the land, is now Ulrich's partner in the aquaponic project. The climate is Mediterranean, with summer months that can be very dry and hot. At the same time, Cajiz lies in a coastal area where fish is traditionally a substantial part of the common diet and as such is consumed abundantly. A strong driver for Huerto Lazo to start an aquaponic project was the awareness that fish is consumed at a higher rate than its capacity to replenish stocks. This is a global trend, although the particular climate and eating culture of the area make the risk of water scarcity and excessive fish consumption more evident and compelling issues.

Ulrich's interest in aquaponics started years ago, when, once he had moved to his current house from Germany, he designed and assembled a small prototype and subsequently a larger one, with a trial-and-error approach, using components available on the market, not specifically designed for aquaponic systems. The initial prototype is still functioning, located in the veranda of the house. It is a small glass fish tank, approximately 150 × 80 × 80 cm, with a small hydroponic unit on top, composed of a PVC horizontal tube and three columns (i.e., vertical tubes) connected to it (Fig. 6.3), forming a small vertical farming system. The water circulates from the tank to the hydroponic unit and back. Each column has holes in it, shaped to offer ledges where blocks with plants are inserted. Ledges are filled with expanded clay pebbles and the irrigation system is a simple drip tube activated by a small pump, utilising water coming from the fish tank. The water drips into the vertical tubes, through the expanded clay pebbles, reaching the horizontal pipe, and finally returning to the fish tank. The filtering system is a small sponge of about 300 cc connected to the pump, which must be frequently changed and that prevents the tubes filled with clay pebbles from clogging. The joints of each segment of the pipes are sealed with silicon glue. This prototype soon presented issues; the glue was not

Fig. 6.3 First prototypes of aquaponic unit

completely sealing the joints and the water dripping through the gaps stimulated mould growth which was damaging plants. Another issue that soon became evident was the unsuitability of the local climate for outdoor aquaponics. Andalusia is a southern region of Spain with hot and dry summers and winter temperatures that exceptionally drop below zero at night. This required the addition in the tank of a small heating unit. But with temperatures above 30 °C in summer, the water in the tank reached relatively high temperatures leading to a slower intake of oxygen in the water and a life-threatening risk to the fish.

The second prototype used a similar fish tank, but the pipe system was not glued, and joints were sealed with rubber rings, which resulted in higher efficiency in terms of water leaks and better health of plants. It was also designed to be easily assembled and disassembled. In this second prototype, techniques for breeding fish were tested and soon the tank was used as a hatchery. The prototype was used to experiment with different types of fish, starting with Tilapia, which is still the most common species in their current aquaponic system, and subsequently adding perk, catfish and even Australian crabs. Catfish proved to be extremely resistant to low levels of oxygen (it can take in oxygen when it is out of water, directly from the air) and growing to big sizes, therefore enabling higher yields. With the experience gained from these prototypes, Ulrich Eich and his partner built a greenhouse where the fish could survive over the cold months and be protected from extreme temperature over the summer months, using natural ventilation (opening panels in the roof) and a sun-shading system. The greenhouse was designed to store solar gains in the water tanks, although gas heaters were also used in the coldest nights of the winter months.

The greenhouse (about 10 m × 4 m) was self-built at low cost, on a concrete floor, with a steel structure and an envelope of polycarbonate panels. It was equipped with eleven plastic tanks, each one with the capacity of 1000 L and containing 100 tilapias, together with a tank with filters. On top of each tank, the column system tested on the original prototype is deployed (Fig. 6.4). At the time Huerto Lazo was visited, the only plant grown in the columns was Jiaogulan, a plant with medicinal properties, regulating blood pressure. Another plant grown by Huerto Lazo is Bocopo Manieri (aka Bhrami), a plant with medicinal properties that, Ulrich Eich claims, helps significantly in slowing down effects of dementia. These crops are particularly appropriate for the local climate.

Next to the greenhouse, there is a third aquaponic unit, built under a much larger shed/greenhouse, which, similar to the first greenhouse, was assembled inhouse with a structure of steel and polycarbonate panels. This facility is the largest, measuring about 50 × 9 m and containing a long pool, about 4.5 m wide and 30 m long (Fig. 6.5a, b). The roof was painted white, a system that is adopted locally to protect from overheating. The shed was built on a water pool constructed about 200 years ago with stones and earth, now populated with tilapia and catfish. The water pool was divided into compartments and Ulrich and Amancio built cages that sit on the

Fig. 6.4 The first greenhouse with jiaogulan plants growing above the fish tanks

Fig. 6.5 (**a**, **b**) Second greenhouse; compartments of the pool with steel frame baskets

edges of each compartment, which are used to catch the fish. Some of the water is used to feed the hydroponic columns, but the aquaculture component of the system is largely underutilised: the water contained in the pool could feed a larger hydroponic unit. Attached to this greenhouse, Ulrich is building an extension, not to expand the greenhouse but rather to design and test a modular and collapsible structure in steel that can be easily assembled as a greenhouse for aquaponics. The aim is to design an affordable system, appropriate for arid climates with scarcity of water. Altogether, the units require daily and methodical assistance as well as periodical maintenance, a full-time job led by Ulrich and Amancio, supported by the young guests of the house.

Outcomes Herto Lazo started as an educational project, with the added aim to generate some income from its produce. The project is very successful in achieving its educational aim. Its outreach goes beyond the local communities and the German young adults hosted, in that they have been able to network with local organisations and propose themselves as a training facility and a demonstration unit. For example, they are connected with Aula del Mar in Malaga, a local organisation that works on sustainability-related projects locally and that organises aquaponics training courses, financed by the Junta de Andalucia (i.e., the local regional authority), the Spanish Minister of the Environment and the European Commission. As part of these courses, participants visit Huerto Lazo and learn about techniques to construct and run systems at a small to medium scale. Because of the self-build and experimental character of the project, participants are introduced to low-cost options to implement aquaponic farms. Huerto Lazo demonstrates that people with no formal training or specific science-based education can grasp the complex cycles of fish breeding and creatively engineer components and systems to farm fish. Moreover, Ulrich Eich designed, tested and built components of the soil-less units, including the columns for the hydroponic units, and filters, which are more affordable than the industrial products and, Ulrich Elch claims, equally efficient. By building independently an aquaponic systems Huerto Lazo have established themselves as expert in this field and as such have been asked by Aula del Mar to contribute to the drafting of aquaponics guidelines.

Some income is generated by selling fish to local restaurants. El Sollo, a restaurant with a Michelin star, is one of their clients, purchasing Tilapia, which is all fed with organic food based on 80% wheat. However, the main source of income is from the fees for hosting young people.

Productivity As mentioned above, fish farmed is sold to a local restaurant, but crops are not commercialised. Huerto Lazo believes that they could not farm sufficient fish to meet the overall demand of the restaurant, let alone the demand from other clients, should they seek to expand their activities. Fish is farmed to attain the highest quality that the client demands since accelerated fattening does not create a good meat texture. The system needs a minimum of 1 year to achieve 400 grams in weight per fish, each of which is sold at €5. The 2019 estimated income from fish sales is €6000. One of the reasons for farming catfish is to increase yields through fish that reach a larger size (catfish can develop to a size of approximately 1.5 kg) and therefore generate higher income. Once again, the quantity of fish harvested shows that the units are not utilised at their full potential. An estimate of the size of the pool in the largest shed suggests that its size can be 4.50 × 30 × 1 m, with a volume of 135 m^3 and a capacity of 135,000 L. Considering the real capacity and an average fish population, this pool could contain – based on an estimate available on the Huerto Lazo website – 'an average of 6000/8000 fish per cycle and 12,000 plants per cycle (about four times a year)'.

Environmental Performance It is difficult to evaluate the efficiency in resource use of this project since Huerto Lazo does not measure water and electricity consumption on a regular basis. Resource use was estimated by the directors as follows. Monthly energy consumption is 2000 kWh, including irrigation, domestic use and more. Estimated energy consumption for operating the units is 1500 kWh per month. Some of the water used in the units, including irrigation of the fruit trees, is groundwater. But the water in the aquaponic system comes from a spring, not metered. A rough estimate provided by the directors of Huerto Lazo is about 10 m^3 of water per day, reduced to 5 m^3 in summer months. An estimated 25 m^3 of water can be added per day in the large unit, and groundwater must be added to the water from the spring. In winter, the water is not changed. The large aquaponic unit is an open circulation system; some of the water is filtered and used to irrigate the trees but not recirculated. No other data is available on the daily water use and quantities of fish feed used, which, as noted on the project's website, is organic and mainly based on wheat. Yet, the continuous process of experimenting resulted in the identification of some benchmarks in terms of space-efficiency and productivity that Huerto Lazo provides on their website. For example, Huerto Lazo claims that, within their aquaponic system, a surface area of 1.2 m^2 is sufficient to support 100 fish in 1000 L and 160 plants.

Table 6.2 Huerto Lazo: productivity and environmental performance over 1 year

Huerto Lazo	Community-led aquaponic project
Objective	Educational; training facility; commercial
Employees/volunteers	2 directors
Total size of the units	Estimated 490 m^2 (10 m × 4 m + 50 m × 9 m)
Production units and area	Greenhouse 1 (estimated) Fish tanks: 11,000 L Hydroponic unit (size not available) Greenhouse 2 (estimated) Fish pool: 135,000 L Hydroponic units (size not available
Initial estimated investment	N/A
Energy consumption	1500 kWh/month
Water added daily	250 L
Density fish	7 kg/m^3
Density plants	160 plants m^2
Fish harvest	120 kg/year
Fish variety	Tilapia, catfish
Crop harvest	N/A
Crop variety	Jiaogulan, Bocopo Manieri
Fish feed	15 kg/day

Ulrich Eich is quite sceptical of the organic certification, which he sees as a way to increase the market value of produce, rather than to diminish the environmental impact of food production. Instead, he is more concerned with the 'absolute' aims of sustainability, therefore looking at ways to address the challenge of water scarcity and fish depletion, which he believes can also be achieved through aquaponics. Data in the Table 6.2 above are based on rough estimates provided by Huerto Lazo.

6.3 Real Food Wythenshawe/Geodome (Community-Led Aquaponic Project)

Directors: Jacqueline Naraynsingh – Real Food Wythenshawe Programme Manager; Kay Bamford – Real Food Wythenshawe Growing Coordinator.
Location: Wythenshawe, Manchester, UK
Website: https://www.realfoodwythenshawe.com
Visited on: February 2020

Motivations Real Food is an organisation that focuses on improving the health and wellbeing of people in Wythenshawe. Its aim is to change lifestyles and food behaviour, focusing on three key areas; growing, cooking and learning, with a mission to 'engage and excite the people of Wythenshawe in growing and cooking fresh, sustainable food' (Real Food Wythenshawe, 2020). Wythenshawe is a neighbourhood on the outskirts of south Manchester, with a high level of deprivation and a population with a large share of unemployment. It is included in the top 10% of the most health deprived areas in the UK (Index of Multiple Deprivation, 2015). Initially, Real Food mainly focused on educating families and school children to appreciate healthy food and improve their diets in a neighbourhood in which many of the families cannot afford organic fresh food and require knowledge to understand the impact of unhealthy food on their lives. Real Food grows food on different plots, including the garden of the Wythenshawe Campus of Manchester College. The Geodome aquaponic unit that Real Food runs is located on that plot and has a particular history. Urbed, a design office and consultancy based in Manchester specialising in urban sustainability, was asked to convert a vacant office block into a vertical farm in Wythenshawe, as part of the Manchester International Festival in 2011 (Urbed, 2020). The commission was for a concept and a feasibility study, which not only focused on space planning, building systems and structural design but also on the integration of aquaponics, water, waste and other systems. Urbed named the project Alpha Farm and developed it with an ambition to demonstrate that indoor farming could be delivered using less energy than industrial farming. The report that was produced included detailed technical aspects, provided by the consultancy firm Biometrix Water and the University of Cambridge. The conversion of the abandoned office block remained at a study stage. Instead, redirecting the initial project, Real Food Wythenshawe was funded with a Biosphere that could champion aquaponics, demonstrating to students how food is cultivated and farmed and producing food to share with local communities. Together with the in-soil cultivation that occupies most of the plot, the aquaponic unit could introduce visitors to a range of options to grow food sustainably. Real Food developed an educational programme for all local schools, teaching: (1) growing produce in non-traditional environments (2) using fish waste as a natural fertilizer to produce a higher yield; (3) the reduction, reuse, and recycle of organic waste for composting; (4) the reduction of food miles; and (4) alternative growing techniques such as aquaponics and hydroponics.

Real Food had access to funding, part of which was used to purchase a kit to construct a geodome from Natural Spaces, a company producing and selling such kits, also to be used as greenhouses. The Geodome is cladded with frosted polycarbonate and has a timber frame (Fig. 6.6). The Geodome purchased has an 8 m (26 ft) diameter and covers a surface area of approximately 36 m^2, an appropriate size for the plot that Real Food occupies on the grounds of Manchester College. Biometrix, specialised in ecology and engineering, and delivering solutions for water management, biodiversity and ecology of places, helped design the indoor aquaponic unit. The floor of the Geodome is constructed in compacted earth. There are raised platforms around the perimeter of the dome on which two plastic fish tanks rest, feeding six hydroponic beds next to the tanks (Fig. 6.7a). Platforms and tanks occupy three quarters of the circle. A second group of five hydroponic small beds and two fish tanks is located in the centre of the dome (Fig. 6.7b). The planning of the space is not designed to maximise production but rather to allow the easy circulation of people within it.

Fig. 6.6 The Geodome in the grounds of Manchester College

Fig. 6.7 (**a**, **b**) Hydroponic beds by the dome and central unit

Outcomes Unlike other case studies included in this book, the purpose of this project is strictly educational and to support local groups. The Geodome has been extensively used as a demonstrator for students and to grow food, which is either given to local people or used to cook meals to demonstrate how healthy meals can be prepared. Potential drawbacks of soil-less systems such as the use of synthetic nutrients and energy, are not usually explained to visitors, since food education and food poverty issues are the main priority of the association. Yet, as a report from the University of Salford (Hardman et al., n.d.) highlights, the Geodome has a strong impact on the local communities and stands in stark contrast with other small scale commercial soil-less farms that – although rooted in the locality – cannot claim to generate the same impact on customers. A Real Food evaluation report (Real Food, 2017) claims that, in the phase of construction of the Geodome (2013–2015), over 100 students from Manchester College volunteered to work on plumbing, construction and landscaping. As noted, courses are organised in the Geodome, on 'composting, recycling, sustainability, food miles, growing your own food and aquaponics'. As of 2017, 832 students participated in these courses or other activities linked to the Geodome.

Real Food secured further funding and developed a new programme based on promoting cooking and healthy eating. The food grown in their plots is cooked in their kitchen, a space within the indoor market in the local civic centre that is opened twice a week. Visitors are encouraged to taste the food cooked by the Real Food team. They are invited to take printed material with recipes of the dishes they have tasted. Real Food also organises cooking classes. Recipes used in the Kitchen and in these sessions, are intentionally basic and cheap, targeting low-income groups. The report from the University of Salford mentions that Real Food constantly engages with schools and universities. In fact, as with other case studies, this association too has links with academia and is willing to contribute to research projects. This consolidates the evidence that there is growing web of contacts and connections between the volunteering and charity sector, civil society and research institutions.

Productivity Real Food does not keep a record of crops harvested and it is therefore not possible to ascertain productivity for this project, although the chief gardener estimates an overall crop harvest from the garden and the hydroponic unit of around 500 kg per year. Crops grown in the hydroponic unit are mainly leafy vegetables while other crops such as carrots are grown outdoors. The chief gardener estimates that up to 20% of the yearly harvest is grown in the Geodome (100 kg) and tilapia harvested amounts to 50 kg per year. Crops harvested are partially used to cook demonstration meals, the number of cooked meals served can give a broad indication of the quantity of food grown. Real Food claims that over 5 years, 15,579 people ate in the Kitchen or attended these sessions, 15,268 recipes were distributed, and 37,482 more people reached through these initiatives and informed about food matters. Real Food is an organisation employing five people. It relies entirely on funding and volunteers and does not charge for the services it provides to local people. Real Food is run professionally, with a clear division of roles that include: a programme manager, a growing coordinator (who looks after the aquaponic system as well), an administrator, an education coordinator and a cooking and behaviour change coordinator. This specialisation enables the association to act on several fronts, from seeking funds to establishing new partnerships. This helps them in operating under precarious financial conditions.

Environmental Performance The aquaculture component is small; there are 2 × 400 L glass tanks and 2 × 700 L plastic tanks, totalling a water capacity of 2200 L. Tanks contain about 20 catfish, 30 crayfish and 20 carp. Fish stock is chosen

Table 6.3 Geodome: productivity and environmental performance over 1 year (2018–2019)

Real food/Geodome	Community-led aquaponic project
Objective	Educational; self-supply
Employees/volunteers	1 growing coordinator + volunteers
Total size of the greenhouse	36 m^2
Production units and area	Fish tanks: 2200 L Hydroponic unit: 6 m tube/4 m^2
Initial estimated investment	N/A
Energy consumption	N/A
Water added daily	Average 5 L
Density fish	Currently about 30/m^3
Density plants	N/A
Fish harvest	50 kg/year (estimated)
Fish variety	Mozambique tilapia, blue lobster crayfish, pangasius catfish and ghost koi
Crop harvest	100 kg/year (estimated)
Crop variety	Kale, lettuce, pak choi, oregano, basil, spinach, watercress, tomatoes and peppers.
Fish feed / cultivated area	N/A

because of their tolerance of dirty water. The nutrient solution from the tanks feeds a hydroponic surface area of about 6.2 m^2. Grow beds are filled with expanded clay pebbles, using the ebb and flow sub-irrigation system. The water in the glass tanks is not filtered and is pumped directly into the beds. The plants act as a filtration system. The water from the plastic tanks is pumped into a sump tank and then into the beds via an indexing valve. Both systems have some netting covering the pipe to minimise large waste particles but there is no proper filter other than the plants themselves. Water in the system is usually checked once a week, when water levels go down, and about 20 L are added as an average, varying depending on the indoor temperature. No additional synthetic nutrient is added to the nutrient solution. The growing coordinator reports that plants grown are in good health and do not present any symptom of nutrient deficiency. Crops grown include mainly leafy vegetables such as spinach, chard, pak choi, lettuce varieties, watercress, kale and basil. Tomatoes have been grown as well.

No data on energy consumption is available as the Geodome is connected to the energy meter of the College, which pays for the energy bills. As part of their educational programme, students volunteering help keeping fish feed and fish health records, as well as performing weekly water tests to check quality levels which may result in water changes, in order to maintain nutrient/fish health balance. The Geodome is never closed. Because of its polycarbonate envelope, the Geodome functions as a greenhouse, accumulating heat indoors. However, the joints are not well designed/fitted resulting in leaks and draughts, therefore requiring heating for the fish tanks in winter. In summer, a cooling fan helps regulate the temperature in the hottest hours, powered with a PV panel. As issues with the design and use of the system are identified, changes are made. For example, two of the fish tanks were originally built in glass allowing visitors to see how aquaponics functions. However, excessive light is detrimental to fish wellbeing, and it was agreed that this should be rectified by applying opaque film on the glass (Table 6.3).

6.4 UGH/Die Urbanisten (Community-Led Aquaponic Project)

Directors: Rolf Morgenstern; Nils Rehkop
Location: Dortmund, Germany.
WEBSITE: https://dieurbanisten.de/urbanisten-projekt/aquaponik-unionviertel/
Visited on: July 2019

Motivations UGH was born as one of the community engagement projects organised by the collective Die Urbanisten, in Dortmund. Die Urbanisten is a multidisciplinary group of architects, urban designers, engineers and scientists with an ambition to engage in live projects and develop new democratic approaches to utilise public space in cities. The UGH aquaponic system is the brainchild of one of the members of Die Urbanisten, Rolf Morgenstern, who has cultivated an interest in aquaponics and who has been experimenting with this technology before UGH started. Rolf Morgenstern is strongly motivated by environmental concerns; he believes urban agriculture and aquaponics are opportunities to attain urban sustainability and create sustainable enterprises and jobs. UGH is an aquaponic unit located in a greenhouse within the complex Union Gewerbehof, a former industrial site, with an old two-storey brick building, now partially transformed into offices, and a towering, corrugated metal cladded, empty factory painted in green (Fig. 6.8). Die Urbanisten squatted on a patch of land, next to a corner of the factory. That land was never reclaimed by any of the companies occupying office space within the complex. Instead, the initiative raised interest. Soon after its construction, the original greenhouse in glass, aluminium and polycarbonate doubled in size. Both modules, now forming one greenhouse, were donated by a local company, producing and distributing domestic greenhouses. The final greenhouse has a footprint of approximately 4 × 5.3 m. (Fig. 6.9).

The greenhouse is orientated to prevent overheating in summer. It does not operate in winter, but the fish is left in the fish tank, where the water never goes below temperatures unsuitable for the fish. The equipment in the greenhouse includes a 1000 L IBC fish tank, a filter for solid particles and a bio-filter.

with ceramic chips. The water feeds two lines of plastic boxes and a linear trough irrigated with ebb and flow, both resting on another water tank, from where the water is pumped back to the fish tank. Grow boxes and troughs are filled with

Fig. 6.8 UGH – external view

Fig. 6.9 UGH – internal view

expanded clay pebbles. The greenhouse has monitoring devices for pH and EC. At the time of the visit, the soil-less system was not running at full capacity. There were about 25–30 almost fully grown grass carp and tench. In a corner of the greenhouse, three more independent units were functioning, with a fish tank below and a smaller one filled with expanded clay and plants above. These units were prototypes being tested with the aim of commercialising a product for domestic use, simplified in equipment and operation. In fact, the units do not have any filters but work with an oxygenator and a pump only. The prototypes in the greenhouse experiment with different types of pumps, with the intention to develop a product that is noiseless, does not squirt water (oxygenators can do this) yet still provide the right amount of oxygen to the water. They are the first product of Plantastisch,[5] a commercial spin-off of Die Urbanisten, seeking to enter the aquaponic market with aquaponic components and simplified aquaponic units.

The main aim of UGH is to become an educational facility and a demonstrator. Since the project started, several individuals, organisations and schools have visited the greenhouse, informed about this initiative through the website, attending events organised at the greenhouse (e.g., seed and plant swap) or joining the annual flea market which is run within the Union Gewerbehof site. The expertise developed through the project led to consultancies. Die Urbanisten were asked to develop the design and coordinate the construction of an aquaponic unit at the Biology Centre SBZ, a centre promoting life sciences to primary and secondary schools in Dortmund. In 2016, they also helped design and build Urban Space Station, a one-year hydroponic installation for the artist Natalie Jeremijenko,[6] celebrating the environmental restoration of the Emscher Valley[7] where the river had been extremely polluted from coal mining.

Outcomes The educational aim of the project was successfully attained. Also, Die Urbanisten were asked to build aquaponic units in educational centres and organise workshops in schools.[8] To date, Die Urbanisten continues to organise regular tours and workshops in the greenhouse. Similar to other projects documented in this book, this one too is the result of a collaboration between academics, practitioners, laypeople and, in this case, activists. This combination of several urban stakeholders blurs the boundaries between experts and non-experts and enables research and innovation to be developed out of the research institutions. The opposite is also true. Community-led projects are used to experiment with solutions that feed into academic research. Rolf's experience in this project has helped advance academic experiments. Rolf is now teaching at the Department of Agriculture at the South Westfalia University of Applied Sciences (SOEST). There, he built one of the few academic aquaponic research and training facilities in Europe. In his teaching and research, Rolf focuses on the development of viable models for aquaponics. He

[5] www.plantastisch.de

[6] https://www.emscherkunst.de/en/kunstwerk/urban-space-station-2/

[7] https://www.dortmund.de/de/leben_in_dortmund/nordwaerts/start_nordwaerts/index.html

[8] https://dieurbanisten.de/author/rolf/

believes that, from a business perspective and for small-to-medium aquaponic companies, selling fish may not generate sufficient profit because of the small margin. Vegetable produce can generate higher profits if it is out of season (e.g., tomatoes in winter) or if it has a high value. The aquaculture component of aquaponics is therefore smaller in terms of profit and should be designed not to maximise yield but simply to provide nutrient to plants. To date, there are not many aquaponic farms operating in Europe. In this light, Rolf believes that aquaponic production needs to be redesigned, and innovative models developed. For example, aquaponic farms could be organised by renting segments of the hydroponic production to individuals or households, similar to allotments plots in allotment sites, or raised beds in community gardens.

The future of the UGH is uncertain. The land around Union Gewerbehof was bought by a developer and the green factory has been recently demolished. The developer has not consulted the companies who own the low-rise red-brick blocks of offices or the other small businesses (e.g., a bar and a restaurant) within the site.

Productivity and Environmental Performance The greenhouse is energy efficient by design. In winter, the fish is not removed from the tank because it would be difficult to start the farming cycle every year. Due to the orientation and location, electric heaters preventing the water temperature from dropping are used only for a few days per year. In the SOEST facility, fish tanks are equipped with electric heaters and the greenhouse is heated from the district heating system. The greenhouse at Union Gewerbehof could be equipped with LCD lamps to function in winter but this system is energy intensive. SOEST is developing an LCA study to precisely identify inefficiencies in this technology, which can be energy intensive (Cohen et al., 2018). UGH does not sell its produce and donates it to visitors. The project survives because of the commitment of the initiators and the occasional funding attracted.

A record of the resource usage and productivity was never kept, because it was considered to be beyond the scope of the project. Die Urbanisten consider this a drawback. The next phase of the UGH project will include the monitoring of the electricity, water, nutrients added and productivity. In terms of nutrients, some are added to the plants when iron and phosphorus deficiencies manifest. Electricity is used only to operate the monitoring devices, the pump, and the electric heaters when strictly necessary. Rolf Morgenstern is collaborating in the European ProGIreg project,[9] designing a prototype of industrial greenhouse with a passive system of storing heat, which would enable the greenhouse to function in winter months with minimal electricity input. The project is at the concept development stage. The data summarised in the Table 6.4 below is estimated.

[9] https://progireg.eu/

Table 6.4 UGH: productivity and environmental performance over 1 year

UGH	Community-led aquaponic project
Objective	Educational
Employees/volunteers	Two directors
Total size of the greenhouse	21 m^2
Production units and area	Fish tank: 1000 L Hydroponic unit: 5 m^2 hydroponics
Initial estimated investment	N/A
Energy consumption	29 kWh/year (estimated)
Water added daily	N/A
Density fish	20 kg/m^3 (estimated)
Density plants	N/A
Fish harvest	Fish is not harvested
Fish variety	Grass carp and tench
Crop harvest	N/A
Crop variety	Mint, chive, lettuce, tomato, water cress, swiss chard, cucumbers, rosemary and other herbs
Fish feed / cultivated area	N/A

Aquaponic and hydroponic farms

6.5 Mangrovia Scicli (Aquaponic Farm)

Directors: Lorenzo Cannella; Arturo Mannino
Location: Scicli, Ragusa – Italy.
Website: www.mangroviaproject.com
Visited on: June 2020 – conference call

Motivations Lorenzo Cannella and Arturo Mannino started their aquaponic farm in 2017, after years of experience in the aquaculture field. They both have qualifications at undergraduate or master's level, informing their approach to this project that has commercial and environmental ambitions. Their studies on marine sciences made them acutely aware of the dramatic depletion of fish stock and the unsustainability of the fishing industry as well as people's dietary habits. Lorenzo Cannella holds a degree in Environmental Sciences and Marine Sciences conferred by the University of Genoa, followed by 3-months of training at the University of Virgin Islands, one of the institutions that has developed influential research in aquaponics (see Chap. 4). Arturo Mannino holds a degree in Biology and a Master's degree in Aquaculture. After his studies he founded Ittica Siciliana, a company farming freshwater fingerlings, a key supplier in this region, without which Mangrovia Scicli could not exist, due to the lack of freshwater fish nurseries in Southern Italy. Scicli is a small town of 27,000 inhabitants, located in an agricultural area in Southern Sicily, not far from the coast. Historically, on this island, there are strong links both with rural traditions and the Mediterranean Sea, reflected in the local culinary culture in which fish and seafood play an important role. There is therefore no tradition or demand for freshwater fish. Farming freshwater fingerlings and starting an aquaponics enterprise in this context therefore comes with many risks. The rate of exploitation of the fish stock is a recurrent topic in the media and there is an ongoing debate about fish farming as an alternative to conventional fishing. Each year the theoretical quantity of fish available worldwide is consumed earlier than the year before (in 2020, this threshold was reached in April). Yet habits, including unsustainable ones, are difficult to change, and there is diffidence towards foodstuff that is not local and traditionally sourced.

The directors of Mangrovia Scicli see aquaculture as an effective response to the challenge of fish stock depletion, in light of a growing global population and lifestyle changes in developed and developing countries, leading to increased demand for high value foods including meat. Aquaculture already provides about 46% of the fish consumed globally (FAO, 2020a) but the market is at present in the hands of large farms and companies at risk, in the directors' opinion, of repeating the errors of intensive farming and environmentally damaging practices. Intensive fish farming can be highly polluting and utilise drugs to protect against diseases, which can be environmentally damaging and affect the quality of the fish meat. On the one hand, the directors believe, the answer lies in decentralised production and a higher diffusion of small-to-medium aquaponic farms. Smaller farms are less prone to prioritise profit to quality of farming practices and they are closer to consumers. On the other hand, technologies such as aquaponics ensure that the environmental impact of aquaculture is reduced. They believe that aquaponics will greatly expand in the future as a technology based on circular economy, recirculating nutrients and utilising waste as a resource.

Water pollution is a much-debated topic. On the southern coast of Sicily there are a few oil refineries that caused oil spills. Other sources of environmental degradation in water such as micro-plastics are also known to people; food scares and the safety of food consumed are becoming common concerns. It is part of the communication strategy of Mangrovia Scicli to emphasise the safety and quality of the food they supply, especially when compared to food produced through intensive use of synthetic fertilisers and meat-based animal feed. But the development of a narrative and an effective communication strategy is difficult, as it requires explaining the complex cycle combining aquaculture and hydroponics with connected advantages. They rely on the relationship of trust which can be built between a small business locally rooted and the local communities.

The farm is in the peri-urban area of the city, hosted in two existing greenhouses that have been adapted for the aquaponic farm. The aquaculture unit is in a 600 m^2 greenhouse, with an envelope in FET opaque membrane to prevent excessive solar gains, and with four plastic tanks with a capacity of 15 m^3 each (15,000 L). The unit includes a bio-filter and a filtration system. The second greenhouse is 1500 m^2 and contains six hydroponic tables, 100–120 m^2 each, totalling 700 m^2 cultivated with a DWC technique (Fig. 6.10). Neither greenhouse is occupied at full capacity. In fact,

Fig. 6.10 DWC beds in the hydroponic greenhouse at Mangrovia Scicli

Mangrovia Scicli started with two fish tanks and one table, and expanded to its current size only last year. The directors are convinced that this slow rate of expansion leaves space to reflect and plan more accurately the next steps. It also allows for experimentation, monitoring and amendments. The directors are well aware that this approach is viable only for farms of small dimensions, in which risks of technical modifications – and failures – are minimal when compared to large farms. Experimenting on a small farm was also for them a way to learn the management of integrated fish-plant systems with a trial-and-error process. The considerable experience accumulated makes them confident to plan for expansion. The aquaponic farm is entirely self-built. This allowed the initial investment for the equipment to be relatively contained (about €40,000). According to the directors, a professional aquaponic system of the same size could cost three times as much. The fish farmed is Largemouth Bass, which was made available in that region when Ittica Siciliana started farming it.

Outcomes The directors are at present the only people working on Mangrovia Scicli. After 2 years of activity, over the last 8 months, the farm broke even and became profitable. The directors are convinced that a period of 24–36 months is necessary to reach economic viability for a farm of this size. Mangrovia Scicli is a commercial enterprise, but it is essential for its commercial success that the quality and safety of their food is effectively communicated. Their network of clients has expanded gradually. The population of Scicli would not be a sufficiently large market even for a farm of the size of Mangrovia Scicli, selling highly specialised products. In fact, the directors are now trying to expand their outreach to neighbouring towns such as Modica and the city of Ragusa, in an area that is densely urbanised. The average age group of customers is 50–60 years old. The network of customers was initially formed through word-of-mouth, and many returned after their first purchase. The network is alerted weekly via WhatsApp where customers can place an order for fish or crops. They are offered three options: collect the food at a collection point (a fishmonger), collect it from their farm or opt for a home delivery with an extra charge. Directors are also trying to implement a new model called Reko,[10] which was started in Finland. It is based on agreeing an appointment at a specific location, where the food ordered can be picked up. Mangrovia Scicli also have a Facebook page which promotes their products and can be used for sales. They are also in the process of subscribing to the platform Alveare (Beehive), an e-commerce food hub of organic and high-quality local producers selling their products to an online community of customers.

Mangrovia Scicli recently started a consultancy service. They promote their farm on social media and on YouTube,[11] where they uploaded other informational videos in which they clearly articulate the potential for aquaponics to offer a sustainable food alternative and the obligation for a new generation of farmers to embrace this technology. They are aware of the vast number of aquaponics videos uploaded on YouTube, mostly from people who are interested in the technology aspects (e.g., fascinated by the design and construction of the aquaponic system) rather than its potential for sustainable food production. This presence helped promoting them as experts; they received many requests for quotations for small aquaponic units, although, so far, only a few followed up.

Productivity Fish population in each tank is typically lower than 30 kg/m^3. Fingerlings are transferred to the tanks when they are 6 months old. Largemouth Bass reach a weight of approximately 500/600gr after 18–24 months of life in the tanks and Mangrovia Scicli now harvest about 1500 kg fish per year. The crop yield is 100,000–110,000 plants per year. The density of plants varies from 40 (parsley) to 20 per m^2. Crops include chards, lettuce, endive and chicory. Typically, income from crops is higher than that coming from fish sales. On average, crops are sold at €4/kg, which is comparable to the price of organic produce in that region. Although the produce is not certifiable as organic, as noted above, it is however produced using nutrient solution, with the only addition of iron in small quantities, when needed; a negligible quantity when compared to that of the fertiliser used in conventional agriculture. Fish is sold at €12/kg, which is perhaps not competitive if compared with sea fish from aquaculture such as sea bass, which is sold at around €8–9/kg. However, freshwater fish in north Italy regions, where the cost of living is higher, are sold at a comparable price or higher when processed and sold as fillets.

Environmental Performance Mangrovia Scicli started as a farm running with a coupled recirculating system but recently moved to a decoupled system to improve water quality for plants and fish. Due to the selection of crops (i.e., leafy vegetables),

[10] https://urgenci.net/reko-a-winning-concept-in-finland/

[11] https://www.youtube.com/watch?v=OHTy_xyzWg8

Table 6.5 Mangrovia Scicli: productivity and environmental performance over 1 year (2018–2019)

Mangrovia Scicli	Aquaponic farm
Objective	Commercial; consultancy
Employees	Two directors
Total size of the farm	600 m^2 aquaculture +1500 m^2 hydroponic
Production units and area	Fish tanks: 60,000 L Hydroponic unit: 700 m^2
Initial estimated investment of the farm	€40,000
Energy consumption	3.5–4 kWh
Water added daily	1000/2000 L/day
Density fish	30 kg/m^3
Density plants	20 per m^2
Fish harvest	150 0kgh/year
Fish variety	Largemouth bass
Crop harvest	100,000–110,000/year
Crop variety	Chard, lettuce, endive, chicory and other leafy greens
Fish feed/cultivated area	25–50 g/m^2

the nutrient solution is enriched only with iron. Each day, 1000–2000 L of water are added to the fish tanks and transferred into the hydroponic system, and from there absorbed by plants or evaporated. Evaporation from the fish tanks is minimal due to the opaque membrane of the greenhouse where the fish tanks are located. The variation of the water input depends on the season and the air temperature. There is no heating system in the two greenhouses; the local climate has only a few weeks of cold weather per year. This period is shrinking due to the changing climate. As a result, a 'natural' balance is struck over the course of the year between the quantity of fish feed needed (less in winter), the level of nutrient in the water, and demand for water and nutrient from the hydroponic beds. The amount of fish feed utilised (which typically varies between 50 and 100 g/m^2 hydroponic cultivation) can be as low as 25 g/ m^2 in the cold season and certainly lower than the top range of the average quantity in summer. Fish feed has a high protein content due to the fish being carnivorous. The energy used in the greenhouses powers only two pumps, one for each greenhouse, two oxygenators and two evaporation systems, which nebulise water in the hottest months. The estimated energy consumption varies between 3 and 4 kWh/m^2 (Table 6.5).

6.6 BioAqua Farm (Aquaponic Farm)

Director: Antonio Paladino
Location: Wedmore, UK
Website: www.bioaquafarm.co.uk
Visited in: March 2020

Motivations BioAqua Farm is a small commercial enterprise, owned and directed by Antonio Paladino, a former chef with long-standing experience, especially in the catering sectors. He has worked for five stars hotels, UK government organisations, UK National Health Service, and private clients, in many European countries and beyond. Over these years, his passion for food has never faded. As a professional chef, he has always been particularly aware of the importance of the quality of the raw material for the preparation of good food. When he moved to the UK, he was disappointed by the quality of the food available in the country. Whether having the possibility to purchase expensive foodstuff when working for wealthy clients or buying affordable food from big retailers for less affluent clients, he grew progressively critical of the food available on the market, especially on the affordable end of the spectrum. And if sometimes good raw material was made available at an extra-cost, it raised questions on the fairness of the food system and a business model of agriculture and farming that could not make available healthy food to all. By the same token, in his opinion, the fairness of the farmers market model is debatable; it offers the opportunity to buy organic food directly from the producer, although the high costs do not always match the quality of the produce. In 2009, he travelled to Malaysia and visited an aquaculture farm, which attracted his interest and pushed him to read more about fish farming. This curiosity and the knowledge subsequently acquired by extensively studying textbooks on soil-less technologies, aquaponics in particular, convinced him to change profession and become a farmer. In fact, he could see that indoor soil-less farming and the possibility to increase productivity by closely controlling crop and fish cycles was the possible answer to the provision of affordable, healthy and possibly organic (or at least grown with organic nutrients) foodstuff. Antonio studied both the technical and financial aspects, looking at the size of the aquaponic farms recently established in Asia, such as the mega-farm recently planned in Xiamen, China,[12] and came to the conclusion that a small-scale farm was financially viable, certainly more viable than a larger one. He started experimenting with aquaponics in the backyard of his house in London with the aim of carrying out a financial analysis and investigating not only the feasibility of the engineering aspects of construction but also the flavours of the food grown and farmed, which could meet his expectations. Following this 1-year phase, in 2010, he decided to apply for a grant under the European and Maritime Fishery Fund, which could fund up to 40% the cost of the system. With that money and a loan, he had access to £100,000, which paid for land, materials, equipment and a website. Labour cost was not included because BioAqua Farm is entirely self-built.

BioAqua Farm is in Wedmore, Somerset, where Antonio relocated. The 1-acre farm is in a peri-urban area, surrounded by many other traditional peri-urban farms. The rural character of the landscape around BioAqua Farm is – to an extent – misleading. In fact, the area is not entirely occupied by farmland and farmhouses, but

---

[12] https://www.facebook.com/pg/Byspokesorg-195693505791/photos/?tab=album&album_id=10153096872335792

Fig. 6.11 Inside the hydroponic greenhouse

also consultancies and professional services offices. On this one acre of land, Antonio built two greenhouses in timber and PET membrane. Although well-constructed and functionable, these greenhouses are rather low-tech, with no particular automated features. For example, one of the sides of the greenhouse has panels that can slide down and let air in, which are manually operated when the indoor temperature needs to be reduced. The first greenhouse is 32 m long and 10 m wide (320 m^2), and hosts the hydroponic system, which is deployed in four linear beds. Each line has a low deep water culture bed and a deep one filled with clay pebbles, utilising an ebb and flow technique (Fig. 6.11). There are other small beds and containers connected to the recirculation system, in which Antonio Paladino experiments with nutrient film technique or even with soil. This is part of an ongoing attempt to experiment with different growing techniques and assess their productivity, viability and capability to conform to high quality food standards. Adjacent to the hydroponic greenhouse, on an area of about 200 m^2, three outdoor beds (each one approximately 20 m × 1.5 m) are also irrigated through the recirculating system either with a drip or with an ebb and flow system. These beds are equipped with supports for crops such as tomatoes and are used only in the summer months, whereas the hydroponic greenhouse operates all year round.

The second greenhouse is approximately 15 m × 15 m. It is timber framed, and covered with shade netting and a roof made of frosted polycarbonate. In the greenhouse there are seven tanks half-sunk in the ground in order to retain heat (Fig. 6.12). Four of the large 10,000 L tanks contain fish or fingerlings and the other three are used for water purification and precipitation of solid waste. Tanks are self-built in zinc sheeting and lined with a mechanically strong, sealed waterproof membrane. Two other plastic containers are sunk in the soil and hermetically sealed with caps: a bio-filter and UV light purifier. Sensors detect temperature, pH and concentration

Fig. 6.12 Inside the aquaponic greenhouse

of solid particles in the water. A sensor is connected to a switch that sends a text to the manager's mobile when the water is below a critical level. Although using sensor technology, the system is not computer operated. There is no automatic feeder: fish are fed manually with frequency varying according to the season. The system is decouple-able, enabling the separation of the hydroponic loop components from the aquaculture ones, thus reducing risks of failure due to an inappropriate water quality either for the fish or plants. The simplicity of the construction techniques, the self-build nature of the aquaponic system, its engineering and the reduction of the electronic components to the minimum necessary for running the plant, contribute to keeping maintenance costs low. Antonio has acquired technical and biological knowledge overtime, enabling him to undertake basic tasks such as monthly cleaning of the tank in which solid waste is separated, monthly cleaning of filters and, when necessary, calibrating the sensors that control water level and the other parameters. A third shed – prefabricated – is positioned next to the farm gate and is used for processing and packaging the food. It is equipped with a refrigerator, two sinks and working surfaces. Apart from an orchard with bushes for berries and some apple, pear and medlar trees, which are harvested, and fruit sold as part of the farm's offer, the land is not used, and the farm has space to expand.

Outcomes This is a commercial farm, designed to be small and to demonstrate that financial viability for this scale of farming is possible. This is because of the high value and fast growth cycle of the crops cultivated as well as the added value of the processed food sold, which includes, for example, pesto. Trout are rarely sold as raw food but rather smoked, marinated or sold in sauce. This increases the market value from approximately £14 to £36 per kilogram. Crops are sold as salad bags or units, mainly through Food Hub, a local network connecting producers with customers and organising collection and sale at some selected locations. BioAqua Farm, for example, delivers its produce weekly to a local collection point in Frome, a few miles from the farm. However, BioAqua Farm has trialled several ways of marketing and selling its produce, including farmers markets and food stalls selling fish burgers. Although difficult to establish, the impression is that this value-added food is essential to the financial success of the farm. In this respect, the director of the farm was able to effectively apply his culinary skills to develop a range of processed products that can attract local customers. However, the next step for BioAqua Farm is to further diversify its offer and include services, rather than grow in scale and production. The director offers consultancies and organises training courses on the farm which allow him to utilise his considerable experience and promote aquaponics, but which also reflect the economic challenges that a small farm faces, requiring creativity in designing a business model with multiple revenue streams. To date, although BioAqua Farm cannot disclose revenue generated from its activities, sales pay for 1.5 FTE farm staff.

Productivity The aquaponic farm was built over 1 year and, when it started, it was not operating at full capacity. The manager went through several iterations before establishing the correct procedures for the aquaculture systems, identifying the suitable crops to grow. Over the year, about 50 different crops are grown, selected for their market value in order to maximise financial returns. Crops such as rocket, coriander, sorrel and samphire are sold for green salads; other crops such as tomatoes, cucumbers and peppers are also cultivated, together with leeks, broccoli sprouts and cavolo nero. The quantity of crops grown annually is not available, but the number of trout harvested every year is approximately 3500, totalling about 1.2 t, averaging over 20 kg of fish every m^3. The farm is harvesting across three short growing cycles, approximately April to June, July to November and December to April. Seedlings are grown in a dedicated heated greenhouse with seeds that are generally bought and only a few coming from the crops cultivated. According to Antonio Paladino, planning crop selection and ensuring wide crop diversity reduces risks for pests. In fact, BioAqua Farm does not apply any particular pest control technique, including netting, but rather relies on the good health of plants, which makes them resistant to any disease. It is a strategy that has not failed him so far. Trout is the only species farmed in the tanks, with stock replenished every 2 months and fish harvested when 8-months old, with a weight of about 300/500 g. Considering the size of the two components of the aquaponic system and a surface area for crops that is 50% of the entire area used for indoor and outdoor hydroponic cultivation, 200 L of enriched water are used for each m^2 of the hydroponic system (Table 6.6).

Environmental Performance The farm was designed using passive design principles to save energy. Energy consumption totals approximately 3 kW/h which are used to operate pumps and the other control devices connected to the fish tanks. The energy supply comes from a provider of renewable energy. The farm runs all year round and the water in the fish tanks never goes below 4 °C; in the greenhouse, heat is transferred from the air and stored in the soil around the tanks and the water in the

Table 6.6 BioAqua Farm: productivity and environmental performance over 1 year (2018–2019)

Bioaqua farm	Commercial aquaponic farm
Objective	Commercial; consultancy
Employees/ volunteers	1 director
Total size of the farm	4000m^2
Production units and area	Fish tanks: 40,000 L Hydroponic unit: 300m^2 indoors +200m^2 outdoors
Initial estimated investment of the farm	100 K
Energy consumption	26,280 kWh/y
Water added daily	142 L daily avg.
Density fish	20Kg × m^3
Density plants	Higher than soil farming
Fish harvest	1200 kg/year
Fish variety	Trout (sturgeons and perch to be introduced)
Crop harvest	N/A
Crop variety	Rocket, shiso, amaranth, parsley, coriander, basil (5 variety), spring onion, chard, samphire, capsicum peppers, chillies, tomatoes (6 variety), cucumber, aubergine, cabbages, kale, pak choi, mibuna, celery, courgette, grapes, fennel, salad (9 varieties), dwarf beans, green peas, mange tout, corn, parsnips, carrots, celeriac
Fish feed/ cultivated area	40 g/m^2

tanks. Furthermore, the speed of the water flow in the piping system is used to regulate heat by, for example, running faster during summer months and slower when it is cold. The rate or water replenishment changes according to the indoor temperature – therefore, the activity of the plants. When evaporation and evapotranspiration are at the highest rate, the system can consume 4000 L weekly, which is roughly 8% of its water capacity. No water is added on winter months. Water consumption for crop production is contained through a careful use of the irrigation systems, with drip irrigation sometimes used depending on the crop. The sludge collected from the tank where solid waste is separated is processed in a digester and subsequently used to enrich the soil of the orchard. This gives excellent results in terms of yield and closes the overall loop of the nutrient cycle of the farm, with nothing being wasted. The food grown is not organic, as there is no certification for hydroponic and aquaculture in the UK. However, according to Antonio Paldino, no synthetic nutrient is added to the nutrient solution. Plants have never shown nutrient deficiency of any type. The fish feed used is possibly one of the most expensive on the market and is certified as organic from its Danish manufacturer. It contains no substance contributing to give a colour to the fish meat (trout), which is white.

6.7 Smart Farmers (Aquaponic Research and Development Facility)

Director: Pascal de Bondt
Location: Gent, Belgium
Website: www.smartfarmers.eu
Visited on: July 2019

Motivations Smart Farmers is an enterprise designing and building indoor aquaponic farms. They started as 'Urban Smart Farm', a highly experimental project. Smart Farmers is an innovation-based company; continuous knowledge development is crucial to its success and the facility visited for this case study is where research and development is carried out. At present, Smart Farmers is not a community-led or social enterprise project, although, in its first phase, it generated community engagement. In 2015, Pascal de Bondt started a small aquaponic unit in two containers located in DOK, an area of Gent occupied by artists and people working in the cultural industry,[13] now earmarked for development. The small unit attracted many visitors and acted as a demonstrator. Pascal de Bondt is driven by the belief that aquaponics can address water scarcity and the inefficiency of industrial agriculture, both in terms of water and nutrients. Phosphorus, for example, is a non-renewable element in growing demand by the agricultural industry, causing water eutrophication (Mayer et al., 2016). With resource scarcity increasing costs, resource efficient technologies and nutrient-recovery techniques become increasingly important. The water efficiency of hydroponic culture coupled with the use of nutrients coming from fish excretion offers an invaluable opportunity to grow food with minimal input.

Monitoring technology and environmental control are key for a successful aquaponic system. The first Smart Farmers project was therefore technologically well-equipped. It was partly self-funded with €40,000 and partly enabled through seed funding attracted with a OYA SEEDS crowdfunding project (https://oyaseed.be/). OYA SEEDS is a seed capital fund that invests in projects that can generate social impact. The programme of the aquaponic unit in DOK included a weekly community engagement event. Every Saturday, Pascal de Bondt organised guided tours in his two containers, which were always attended by an average of 20 people and served as a test ground (and a marketing exercise) to probe public receptiveness to this farming system and its produce, which was generally positive. One of the two containers hosted the aquaculture component and the second the hydroponic unit. In the former, Tilapia was farmed in two 1000 L water tanks. The system was complemented with a filter to separate solid waste and a biofilter to break down ammonia.

In 2017, Pascal de Bondt was awarded a grant from the Belgian Technology Board (around £25,000), enabling him to expand the initial aquaponic unit. The DOK community had been evicted to enable the development of the site. De Bondt

[13] www.dok.gent.be

Fig. 6.13 Hydroponic section

identified an alternative site, north of the city centre, adjacent to a wastewater treatment facility, where the farm could relocate and expand. The use of nutrient rich water from the facility became an opportunity for Pascal to attempt to go one step further in his experiment by testing circular economy food production. The municipality subcontracts the wastewater treatment to a private company although it retains control over the financial management and the overall policy. The first challenge was to convince the company and connect their water outlet with the aquaponic unit. The attempt was to test an aquaponic farm connected with the wastewater plant, with the company as a partner and investor. The company showed interest in the project. Currently, Pascal de Bondt is planning to submit the project to the Belgian Technology Strategy Board and seek funds for an experimental unit.

In the current arrangement of the Smart Farmers farm, the first of four containers hosts three fish tanks (totalling 4000 L) (Fig. 6.13). A fourth tank (1000 l) is filled with water coming from the wastewater plant. The water quality in both groups of tanks is constantly monitored. The tanks are populated at 50% of their maximum capacity (i.e., 50 kg/m^3) with red tilapia. Tilapia grown in each compartment will then be tested to identify substances or pathogens that accumulate in their flesh and verify their suitability for consumption. Tilapia is an ideal fish for the experiment and a very common fish in aquaponic farms, but Pascal believes that this is not the most commercially viable option. Trout, for example, would allow higher returns. The system is completed with two tanks with filters. The second container hosts the

Fig. 6.14 Aquaculture section

hydroponic unit (Fig. 6.14). It is divided into three compartments, reproducing three different environmental conditions in terms of temperature and air humidity, each one growing crops that are fit for that particular environment. For example, tomatoes are grown in the compartment with the whitest light. In another compartment, conditions are such that Salicornia can be grown. The plant is generally foraged. As such, it is a high-value crop that is at present particularly in demand by restaurants. Air is exchanged between these two containers, transferring the air from the aquaponic container, with higher CO_2 concentration, to the hydroponic one, in order to ascertain benefits for plant growth. Water in the system is replenished at a rate of about 5% a day. This input can vary, depending on how rich the water is with nutrients, leading to a lower intake of water. Therefore the composition of the fish feed is key to keeping the system in balance, together with the temperature which can determine higher or lower levels of nutrient intake.

The third container hosts the filters, together with a water tank where an experiment of mineralisation is taking place. Pascal de Bondt is experimenting with the solid waste as a by-product of the filtered wastewater. This solid waste is separated and subsequently put into a water tank to release its nutrients. If the experiment is successful, a lower additional synthetic nutrient input for plants may be required.

Outdoors, three hydroponic ebb and flow grow beds are deployed in order to experiment with struvite, the solid waste produced by the wastewater treatment plant. Each table has four plants that are irrigated with water with different concentrations of struvite. Finally, the fourth container houses four tanks with freshwater shrimps. This is a high-value food, especially when compared with red Tilapia which is sold at a retail price of €10–11/kg, and much less if bought frozen. Pascal de Bondt filled the 1000 L water tank with 500 larvae, transferring shrimps to other containers when they grow in size. Shrimps, however, are likely to enrich water with a mix of nutrients that is different from that produced by Tilapia or other fish. The next step is to experiment with this different nutrient solution with crops. A hatchery is also installed in order to make the entire farming cycle self-sufficient. Pascal is planning to build a vertical farming unit over each water tank, in order to optimise the use of space.

Outcomes This project is driven by the entrepreneurial skills of Pascal de Bondt and his attempt to demonstrate the value of waste for soil-less technologies. Unlike other projects presented here, Smart Farmers seeks strategic partnerships enabling investments and gaining credibility for the experiment. Pascal de Bondt is fully aware that the aquaponic market is challenging and with low profit margins. This is also true for the entire agricultural and farming sector, particularly for staple foods. High-tech food production, with its high capital costs, is viable at an appropriate scale and with correct strategies, including marketing strategies. Integrating and advancing organic waste management is therefore important to reduce costs by identifying possible synergies with other sectors. In a future scenario characterised by resource scarcity and the negative impact of waste on the environment, the economic advantages of this experiment can be substantial. Against this backdrop, Smart Farmers evolved from a small farm to a knowledge and value driven small commercial enterprise, developing, marketing and building indoor farms. Experiments in nutrient recovery, different concentrations of nutrient solution, air exchange and more, are valuable for the research into new solutions within the aquaponic sector. These experiments are developed by a small team (Pascal de Bondt, a partner specialised in bioengineering and an intern) and within a relatively low-cost facility. In this way, Smart Farmers is successful in developing new knowledge that is not only marketable but is also produced out of the industrial research and development sector. Smart Farmers acts as a low-cost, flexible, highly independent and creative laboratory, demonstrating that that this type of activity can be carried out by anyone who has strong motivation and interest.

Productivity and Environmental Performance Production and environmental performance are key factors for the sustainability of aquaponic technology. Smart Farmers, as noted, develops solutions on this front too. However, the company is not organised to maximise production and performance but to test them. Optimisation of artificial lighting per crop, can improve productivity. Testing options is energy intensive; in fact, Smart Farmers utilises 31,000 kWh/year for the hydroponic unit only, equivalent to 738 kWh/year/m^2. The daily water input is 117 L, roughly 16% of the water tank capacity on a weekly basis. The fish and crop harvests are estimates as crops and fish utilised in the experiments vary greatly. For example, Smart Farmers believes that their facility can grow 5 T (104 kg/m^2) of leafy greens per year

Table 6.7 Smart farmers: productivity and environmental performance over 1 year (2018–2019)

Smart farmers	Experimental aquaponic farm
Objective	Consultancy; research and development
Employees/volunteers	1 director +1 partner +1 intern
Total size of the farm	4 × 42 m^2 containers = 168 m^2
Production units and area	Fish tanks: 6000 L Hydroponic unit: 48 m^2
Initial estimated investment of the farm	N/A
Energy consumption	32,300 kWh/year PV panels generating 2 kW peak
Water added daily	117 L/day purified waste-water (43.000 L per year)
Density fish	50 kg/m^3
Density plants	N/A
Fish harvest	600 kg/ year
Fish variety	N/A
Crop harvest	4,5MT baby leaves/year
Crop variety	N/A
Fish feed/cultivated area	720 kg feed/year

and 800 kg/ year of fish (133 kg/m^3). These figures, provided by Smart Farmers, are probably aspirational. Data provided here are therefore not always relevant for a comparative analysis with the other case studies. A summary of the characteristics, productivity and environmental performance of the farm is shown above (Table 6.7).

6.8 GROWx (Hydroponic Vertical Farm)

Director: Ard van de Kreeke
Location: Amsterdam Netherlands
Website: www.growx.co
Visited on: July 2019

Motivations GROWx is located near Amsterdam Arena, a commercial district in south Amsterdam, partly populated by high-rise office buildings, headquarters to some of the most high-profile global companies, and partly by low-rise warehouses and offices. GROWx shares one of these warehouses with a company renting bicycles, which takes up most of the internal space. GROWx is owned by a green investment fund, which started the company in 2015 and hired a manager to run it, with the agreement to share profits equally. The manager did not attain production and sale targets. One of the reasons for this was connected to the selection of crops, the value of which was too low to generate sufficient profits. Eventually, the manager left after a few attempts to make the company profitable. Ard van de Kreeke took over this position in April 2019, and in his opinion, he is on target to meet commercial goals.

Before relaunching this company, van de Kreeke had a small farm in the south of the Netherlands which he had to leave because it was not sufficiently profitable. He started his farming business by growing local and seasonal crops, with a high-quality restaurant in Amsterdam as his main client, which soon became his only client, buying his entire harvest. The particular crops and quantities demanded by the restaurant, mainly high-value crops such as lettuces, micro-greens and other leafy greens, could be produced only with indoor farming. Ard van de Kreeke promptly changed farming techniques accordingly, concentrating his efforts in the greenhouse of his farm, where it was possible to yield multiple harvests per year. Moving the production from conventional crops to crops such as micro-greens required organisational changes. His small greenhouse (about 20 m^2) was no longer sufficient to meet the demands of his client; he looked for a larger greenhouse, equipped for the type of cultivation the business required. Ultimately, he decided to become the manager of GROWx, bringing to this company his client, and believing that hydroponics was the most appropriate technology to match supply with demand. Furthermore, the location of GROWx was closer to the restaurant, and this would reduce transport costs and food miles.

GROWx has a very well structured production system. It occupies only a portion of a medium-sized warehouse. The space is organised in functional areas. There is a cleaning station where PVC channels (troughs) are sterilised after the harvest, with workbenches to clean and repair equipment, and plant seeds in blocks and trays. The flat bottom of the channels is covered with a strip of a cellulose-based material and seeds are sprinkled on top. Next to this area, there is a 20 feet container (6.09 × 2.49 m) that is used as propagator, equipped with a single electric heather and a humidifier for basic indoor environmental control (humidity at about 95%). Germination takes 5–10 days, depending on the crop. Channels with seedlings are transferred onto other racks and moved to the next area, which is a thermally insulated cell of approximately 100 m^2 (Fig. 6.15a). The cell is hermetically sealed and built as a refrigerated room. Inside, racks in stainless steel are connected with the irrigation system which floods the channels twice a day, saturating the cellulose mats (Fig. 6.15b). Each rack is about 2 m high, with four levels, each one equipped

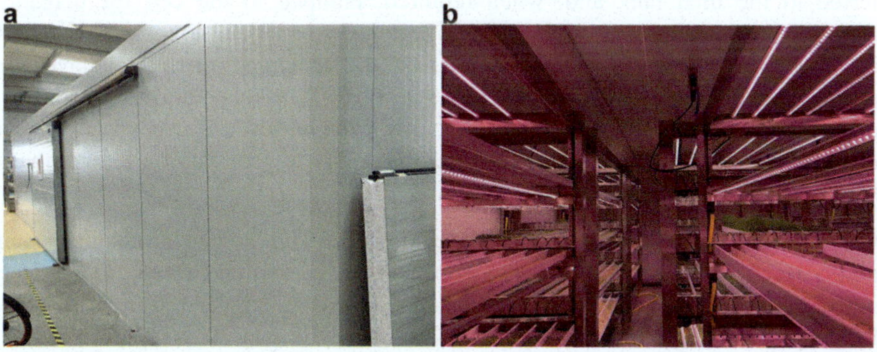

Fig. 6.15 (**a, b**) External and internal views of the growing room

Fig. 6.16 Water tanks and nutrient mixing system

with LED lamps. LED colours are calibrated for each particular crops produced by the hydroponic farm. Lighting time, irrigation cycles, humidity and temperature are controlled by a computer and monitored by a dedicated software program. Crops grown include mustard, chives, mung beans, chickpeas, broccoli, radish and coriander, with a growing time, including germination, varying between 10 and 20 days. Air is exchanged to provide sufficient levels of oxygen; the air-conditioning unit is positioned on the roof of the warehouse.

The last area is occupied by three large water tanks. The largest contains the water returning from the hydroponic cell. In this tank, the water is filtered and treated before being recirculated. This water is transferred to the second tank where it is mixed with two different, complementary nutrients. The water is finally transferred to the final tank and, when required, pumped to the cell for irrigation (Fig. 6.16). A computer directs all phases and monitors indoor temperature, humidity, Ph, CE and more. The nutrient dosage and the schedule for irrigation have been accurately studied and the software ensures a correct nutrient solution. The software also works as a data storage enabling records to be retrieved if necessary. A small office of about 15 m^2 completes the operational space occupied by the company, which totals approximately 400m^2.

Outcomes The main motivation of this enterprise is commercial, although sustainability is a driver and also a component of its communication strategy. On its website, GROWx is presented as a 'fully robotized, zero waste, vertical farm, powered by green electricity'; 'a future proof solution for the way we supply cities with food'. GROWx is an urban farm, therefore producing next to where the food is consumed; it is powered only by renewable energy (although it is not specified whether this is generated from PV panels on the roof of the warehouse or simply supplied by a renewable energy company); it uses recyclable packaging (described as circular packaging); and - of course – operates with technologies that save water, space and do not require pesticides. Unlike the other case studies, production is entirely computer managed, relying strongly on digital technologies to steer towards sustainable food production. Another outcome is success in business, which if confirmed over the following years, is relatively uncommon. GROWx is now in the process of expanding, after an initial period in which operational, technological and market-related issues had to be resolved. The expansion is possible because of a growing network of clients, which started with one restaurant and expanded through word of mouth, and participation in commercial trade fairs and events on food technology. GROWx now supplies 70 restaurants nationwide. The hydroponic unit is running at 70% of its capacity but soon will reach full capacity. GROWx is planning to rent a space in Eindhoven and, if business expands further, one in Rotterdam. This will further reduce the impact of transport since each production unit will provide food to restaurants within their city/district. GROWx is also experimenting with new crops, in three rooms located on the first floor of the warehouse. Some of the restaurants they supply would be interested in edible flowers, which are high value crops: a punnet of edible flowers can be sold for €5. Tests focus on the duration of growth cycles, optimal temperature, humidity and lighting times (some flower plants will flower only when daylength is at least 11 h).

Productivity The GROWx business model is based on a fast growth cycle and on high value crops. GROWx claims that, after 4 months of production, the company generates returns. However, these must be quite small. GROWx produces 1000 punnets weekly (containing an estimated 50 g of micro-green each). The monthly turnover is €20,000. At the time of the visit, there were only two internships working on the hydroponic system, who although not paid are learning new skills. Seeds are costly, but these costs are relatively low when compared to salaries and energy bills. The plan to expand with other indoor vertical farms in other cities seem to suggest that van de Kreeke is fairly sure of the validity of his business model and that the investment group owing the company is behind him. In 2020, 1 year after GROWx was visited, the company was planning to double the growing area (140 to 300 m^2) and become soon after very profitable.

Environmental Performance Although small, GROWx operates with sophisticated equipment and environmental control systems. It utilises approximately 30 kW/h, with 19 kW/h only for lighting, 9 kW/h for the environmental control systems and the remaining 2 kW/h for all other equipment and appliances. Nevertheless, a lower level of energy consumption is one of the objectives of the company. LED are efficient but current lamps convert only part of the power into light and the rest is dissipated as heat. Philips has just issued a new LED lamp which saves 30% energy compared to the previous LED type. Van de Kreeke is pressing on the investment fund owning GROWx, trying to convince them to substitute the

Table 6.8 GROWx: productivity and environmental performance over 1 year (2019–2020)

GROWx	Hydroponic vertical farm
Objective	Commercial
Employees/volunteers	1 director + 2 interns
Total size of the farm	About 400 m^2 (estimated)
Production units and area	100 m^2 (estimated)
Initial estimated investment of the farm	N/A
Energy consumption	300,000 kWh/y
Water consumption	a few $m^{3/}$ month
Density plants	N/A
Types of crops	Microgreens and leafy greens
Crop harvest (kg/day or week)	1000 punnets a week

entire stock of lamps with the new model by Philips. There is also particular attention paid to the use of materials. The cellulose strip was chosen not only for its practicality but also because it is fully recyclable if compared to other materials such as rockwool. Micro-greens are delivered in plastic boxes that are fully recyclable and that the restaurant must return. These are washed and reused regularly. Boxes are delivered to restaurants with the company's electric van in order to minimise the use of fossil fuel. The company is fully aware of the water efficiency of hydroponics, is convinced that their production is sustainable and is committed to further reducing the use of resources. When asked about the reasons for this commitment, the intern who guided the visit responded by emphasising the commitment to sustainability of the entire country, therefore of all people and businesses (Table 6.8).

6.9 Farms That Closed

6.9.1 Unit 84/GrowUp Urban Farms (Aquaponic Farm)

Directors: Kate Hofman; Tom Webster
Location: London, UK
Website: www.growup.org.uk
Visited on: 2018

Motivations GrowUp Urban Farms is a London-based enterprise committed to "producing food in cities at a commercial-scale in an ecologically sustainable way". It operated indoors, in a warehouse, with an integrated system of aquaculture and vertical hydroponics of considerable size. GrowUp Urban Farms started in 2012 with a demonstrator: a small aquaponic unit composed of a greenhouse on top of a container, called The GrowUp Box. The demonstrator was funded through a crowdfunding campaign and was located in a playground in between two schools in South

Fig. 6.17 GrowUp box: a prototype

London (Fig. 6.17). The container had a fish tank inside, connected to the Zipgrow towers in the greenhouse above. This demonstrator was a temporary project, useful for the founders of GrowUp to familiarise themselves with this soil-less technology and test the openness of the visitors and local restaurants to food grown with it. A grant from Innovate UK (a UK national research and innovation council) and an investment from Ignite Social Enterprise, enabled the implementation of a farm in 2015, called Unit 84 (Figs. 6.18 and 6.19). Unit 84 employed 12 people before closing in 2018.

They were the first aquaponic enterprise in Europe completely indoors, using no natural light. The enterprise tested a business model that is not only energy and resource efficient but also socially responsible. Workers were recruited from a charity that worked closely with the London Borough of Newham (the local authority), targeting low-skilled young people living locally who were not in education, employment or training. Employees were given the opportunity to take external qualifications relating to their work (for example Food Safety and specialist equipment operations) and were paid the London Living Wage. This employment initiative had a strong impact on the way they were perceived by local policymakers and local groups. As a spin-off from the main business, GrowUp Community Farms CIC was created; an organisation advising on the construction of small-scale, community-based systems and carrying out education and outreach activities on sustainable food.

Fig. 6.18 Indoor view of the hydroponic vertical farm

Fig. 6.19 View of the aquaculture farm

Outcomes The farm was built as a commercial prototype, to demonstrate the viability of producing food in an aquaponic system and identifying suitable products and routes to market. The farm generated significant local and national media coverage, as well as running weekly "open day" visits for the public. Through GrowUp Community Farms CIC, the company's partner organisation, the team developed

Table 6.9 Unit 84: productivity and environmental performance over 1 year

UNIT 84	Commercial aquaponic farm
Objective	Commercial
Employees/volunteers	12 employees
Production units and area	Fish tanks: 36,000 L tanks Hydroponic unit: 75 0m^2
Technique hydroponics	Flood and drain benches, vertically stacked
Initial estimated investment of the farm	£750,000
Energy consumption	100% renewable energy from energy company
Water added daily	N/A
Density fish (n/m^3)	66
Density plants (n/m or n/m^2)	N/A
Fish harvest	4000 kg/year
Fish variety	Tilapia
Crop harvest (quantity harvested daily or each week	385 kg/week
Crop variety	Leafy green salads, herbs and microgreens
Fish feed/cultivated area	N/A

educational materials for use across all primary school key stages relating to aquaponics and sustainable food production. Unit 84 produced excellent quality produce with high yields for an indoor aquaponic system. The overall facility was limited in its ability to expand due to lack of additional space and limited processing and packing facilities. For this reason, Unit 84 was not able to sell into large retailers as the processing area could not be accredited as required. The farm operated continuously for more than 2 years, but the decision was made to close it at the start of 2018 due to high overhead costs of running the facility (in particular: labour, energy, and rent). This difficult decision meant that the business could take the lessons learned and knowledge developed at the farm and put together a new business strategy that was designed to address some of the commercial challenges faced at Unit 84. This new strategy included focusing on hydroponic instead of aquaponic farming, so as to be able to create a high care production environment, and also to co-locate all future farms with baseload renewable energy generation.

Productivity At full operations, the system was designed to produce over 4 tonnes of Tilapia and over 20 tonnes of salads and herbs per year (about 200,000 salad bags). During its operations, the farm produced and sold over 75,000 units of microgreens and salads, as well as a steady supply of Tilapia. The aquaculture system comprised of 12 × 3000 L tanks, each capable of holding up to 200 fish. Tilapia were purchased as fry and grown out in the system before being sold into a local Thai restaurant group. The fish were fed on an industry-standard feed, and the wastewater from the aquaculture system was used as the nutrient solution for the hydroponic system.

Environmental Performance GrowUp Urban Farms was fully committed to sustainability. Hydroponic production utilised recycled carpet fabric as a growing media and low energy LED lighting that has been specifically designed to only provide the wavelengths of light that the plants need to grow. No insecticides, pesticides or fungicides were used. Standard hydroponic A/B fertilizer was used, together with standard industry Tilapia feed. All deliveries were done by electric vehicle to local clients across London such as restaurants and retailers. The system was entirely powered by a renewable energy and the packaging was recyclable and recycled. The food was delivered within 24 h from harvesting, resulting in fresher produce that lasted longer. The farm worked in partnership with several local food waste charities to ensure that any surplus produce was used in the community and not thrown away. Details on productivity and resource use were not disclosed by GrowUp, except for very broad data and estimates, summarised in Table 6.9.

6.9.2 Bristol Fish Project (Community-Led Aquaponic Project)

Director: Alice-Marie Archer
Location: Bristol, UK
Website: bristolfish.org
Visited on: 2018

Bristol Fish Project is still an active association although, at present, the farm is closed. In starting this project, Alice-Marie Archer, the director, was motivated by the objective of making cities a better place to live. After her Bachelor's degree in Environmental Sciences, followed by a few jobs in the food sector (from waitress to chef to culinary business owner), she became a researcher at the Schumacher Institute, where she cultivated her interest in food. There, she worked in projects identifying ways to use food as a catalyst of lifestyle change. To this end, she modelled food systems using causal loops in order to identify pathways to their transformation. She also studied resource consumption rates, including depletion of phosphate, which are fundamental to industrial agriculture practices. Waste was another research focus, looking at cities as generators and principal utilisers as well as at many bottom-up, food-growing, practical initiatives exploring waste as a resource, showing how these hands-on projects were ahead of research. Building on her keen interest in community projects and activism, her academic investigations and an attitude to put in practice – rather than speculate on – findings from her studies, she started her first community project. A realisation that the impact of industrial agriculture has not been fully acknowledged and her studies on resource consumption led her to trial aquaponics as a low-input technology of food production.

In 2012, a polytunnel was built in the car park of the Art-Space Life-Space College in Knowle, containing fish tanks with a capacity of 5000 L. Tilapia was grown in the tanks and several salads in the hydroponic units. About 60 volunteers donated their time and energy to build the polytunnel and the project attracted media attention. The project was agreed with the College – although not with the planning department in Bristol City Council – as part of a 5-year initiative in which artists were creating artworks and installations temporarily hosted in this place. Although not a piece of art, the construction of the polytunnel was initially discussed and subsequently welcomed by the artists and functioned for 1 year only.

This first project was not only an occasion to trial aquaponics but also to refine recruitment skills. Volunteers were recruited not only through word of mouth but also social media (Facebook and Twitter), which in 2012 were not as widely used as today. In fact, the idea of an aquaponic farm attracted many people, including those with the right skills and knowledge (which Alice-Marie Archer did not possess yet) for building the soil-less system. In her opinion, a more conventional urban agriculture project would have not attracted as many people. Factors such as the creation of an artificial growing system, control of it and the possibility of observing the development of crops and fish were key to recruiting a wide range of people with diverse skills, including engineering skills.

Following the dismantlement of the polytunnel in 2013, Alice-Marie Archer organised and delivered courses in aquaponics, and designed and implemented some units, sometimes helped by the students on her courses. One of these projects was located in the Phoenix Cafe (central Bristol), which lasted a year only. This is a vegan café which was doubtful about the vegan credentials of vegetables that were grown with animal derived nutrients. Another one was located in a local hydroponic gardening store. The final headquarters of Bristol Fish Project was built with funding attracted from Bristol Fund Capital, a fund organised to fund projects celebrating the nomination of the city as European Green Capital in 2015. The headquarters was in an old, abandoned warehouse, which was repaired and upgraded by volunteers. Since then, the Bristol Fish Project grew steadily, with volunteers building fish tanks that gradually occupied the space. In 2018, at the time of the visit, the association was using the fish tanks for a conservation project that was generating income, funded by the European Marine Fisheries Fund and the Sustainable Eel Group. The project consisted of breeding an endangered species of eels to subsequently release them into their habitat. Other projects that Bristol Fish Project was developing included the testing of a breeding cycle of Black Soldier fly, the larvae of which can be used as feed for fish, fed with locally composted waste food.

Mushroom farms

6.10 RotterZwam (Mushroom Farm)

Directors: Siemen Cox; Mark Slegers
Location: Rotterdam, Netherlands
Websites: www.rotterzwam.nl / www.mushroom-cultivation.com
Visited on: June 2020 – conference call

Motivations RotterZwam was founded by Siemen Cox and Mark Slegers in January 2013. Before starting this enterprise, Siemen Cox studied permaculture, planning to apply permaculture principles on a farm that he was hoping to open in an arid climatic zone. In permaculture, cultivation is designed in a way that each element contributes to form a self-sustaining ecosystem while having multiple functions. Siemen Cox used this concept to shape RotterZwam. While attending the permaculture course, he was particularly impressed by a book that he read: The Blue Economy, in which the author, Gunter Pauli, presents 100 innovations that have the potential to steer the world towards a sustainable course. One of such innovations is a closed loop system of mushroom farming, which utilises coffee grounds as growing media. This was another source of inspiration, in that it promoted waste as a resource and sustainable enterprise as an approach to repurpose economy to make it fit for the sustainability of the planet. Together with his current work partner, he started a small mushroom farm in the basement of an abandoned indoor swimming pool under a circular glasshouse, the Tropicana, located by a canal in Rotterdam. The choice of a disused building for their farm was consistent with the idea of a 'waste', circular economy philosophy underpinning their core business based on spent coffee grounds, which in the Netherlands totals 120 million pounds (54.5 million kg) per year (Cox & Slegers, 2014).

Similar to the stories of farmers in the other case studies, the directors of RotterZwam had to learn technical and organisational aspects of this production by themselves. Experts in this field were contacted but, generally, there was reluctance from their side in sharing knowledge. They firstly produced shiitake mushrooms, using old oak logs and spent coffee grounds. In parallel with these initial attempts, RotterZwam was a catalyst for the reuse of the Tropicana building, which soon became a headquarters for start-ups working in the field of waste and circular economy, clustered under the initiative named Blue City. On the project's website, Blue City is defined as an 'ecosystem of social entrepreneurs and radical disruptors focusing on waste'. Starts-up include enterprises that work with textile waste, or produce rain gear from recycled umbrellas, or consultancies specialised in sustainability assessment processes for enterprises.[14]

In Blue City, RotterZwam had the availability of 1500 m^2, with sufficient room to expand production, should the business grow. In fact, by 2017, it was producing 250 kg of mushrooms per month. The initial investment to buy equipment and set the farm was modest (about €20,000). Profits were reinvested to complete and expand the equipment, and all the necessary materials. Clients were and still are mainly restaurants, allowing a higher return for the company. In 2017, a fire damaged the Tropicana building and the mushroom farm, and RotterZwam had to change location. The two directors launched a crowdfunding initiative that successfully gathered about €400,000. This money paid for the relocation and equipment for a new farm that opened in 2019 (Fig. 6.20). Blue City is still operating in the Tropicana building and RotterZwam is still connected with this initiative, using

[14] www.bluecity.nl/en

Fig. 6.20 View of the new mushroom farm

space for the offices of the company. The new farm is hosted in eight containers, totalling 650 m^2 of surface area, located on the canal, west of the previous location. Today RotterZwam converts into substrate between 6000 and 7000 kilos of spent coffee grounds every month, producing 300–400 kg of oyster mushrooms. With the equipment and space available, at full production, the new farm can harvest between 1200 and 1500 kg per month. The farm could not be visited because of COVID-19 travel restrictions, but a video on YouTube,[15] shows that the farm is organised in two groups of containers, with the biggest hosting the growing room (Fig. 6.21), which is kept at a temperature of approximately 20 °C, and the smallest hosting the preparation of growing bags and other processing phases.

In spite of their entrepreneurial attitude and their constant growth, the two directors distance themselves from the big producers of mushroom farms and their ambition is to keep their farm in an urban context, therefore next to the place of consumption, where waste is generated and can be utilised as a resource. This implies that rather than growing in terms of quantity of mushrooms produced, the company can grow by diversifying their offer and/or replicate the farm in other cities.

[15] https://www.youtube.com/watch?v=4Ue_ZQjQ-4U

Fig. 6.21 Inside one of the containers of the new mushroom farm

Outcomes RotterZwam is today a successful creative and sustainable enterprise. The idea of multifunctionality embedded in each element of an ecosystem, one of the permaculture principles, has inspired Siemen Cox in shaping a circular business model, diversified in products and services offered. To build resilience in their enterprise, other initiatives were added to mushroom farming, which could ensure a sufficient stream of revenues. Today, mushroom farming represents approximately 30% of the total income. Another 30% is generated through the sale of the GrowKit, a kit to grow mushrooms at home, designed by RotterZwam. Another stream of revenue comes from the collection of spent coffee grounds which would be otherwise disposed. Restaurants or any other company committed to sustainability through their Corporate Social Responsibility policy are willing to commission the collection of this waste and to recognise a cost for this service. In the Netherlands, CSR is becoming increasingly important as companies' sustainability policies become one of the criteria required to pitch in public work tenders.

Income is generated also through the sale of vegetarian mushroom snacks such as bitterballen, a normally meat-based national snack, appealing to vegetarian and vegan consumers. Mushrooms are used to replace meat content, being an excellent substitute for flavour and texture. RotterZwam subcontracts the production of these snacks to local caterers and sells them to restaurants. This diversification enabled RotterZwam to survive in the year between the fire and the start of the new farm. Other products include a mushroom beer and a gift pack with their products, which is sold in stores and during educational tours of their farm. This creative diversification is helping the company get through COVID-19 despite the temporary closure of restaurants, which are their biggest customer. The sale of GrowKits has increased and they have been able to redirect their produce to local farmers markets. A new concept that the two directors are developing uses the idea of the service economy, and is based on renting rather than owning, applied to the coffee lifecycle. They are planning to design a platform which puts together coffee roasters, coffee drinkers and coffee 'processors' in order to ensure that all these stakeholders work in line to minimise coffee waste.

Their company has attracted extensive media coverage and increased its outreach through a website, a blog and an active presence on social media. Their website in particular, offers resources to those who are willing to approach mushroom farming, with downloadable open access e-books and articles providing practical advice as well as other resources requiring a fee. The open access e-book '10 things you need to know when cultivating mushrooms' (Rotterzwam, n.d.), states that the ethos of the company is 'share all you have and you will receive more' and presents mushrooms as the substitute for meat. They have a YouTube channel with many resources available on mushroom growing and the design of mushroom farms. A video playlist is also available with short interviews with the graduates from a Master's course that they have designed, which covers technical, entrepreneurial and marketing aspects. These former students are now directing farms across Europe such as Beyond Coffee in Denmark, Hut und Stiel in Austria, Le Champignon de Bruxelles and Helsinki in Finland.[16]

Their dynamism and expertise have generated demonstrable impact through their Master's course and their activity in consultancy in mushroom farming. At the same time, they are keen to reduce their environmental impact by measuring it and reducing it. In fact, they commission yearly reports on their environmental performance, which are shared on their website. Finally, they are committed to developing impact through academic research. With the project Back to the Soil, in partnership with Wageningen University, they are experimenting with techniques for the enrichment of the soil, using spent substrate from the mushroom farm.[17]

Productivity As noted, prior to COVID-19, RotterZwam was producing between 300 and 400 kg oyster mushrooms, selling them to restaurants at about €8.5–10 per kg, against a retail market value of €7. In an article published on Urban Agriculture in 2014 (Cox & Slegers, 2014) the directors state that their commercial objective is to produce 7500–18,000 kg per annum resulting in an income of maximum €162,000, double the production and income seen at present. Ultimately, a policy of diversification was preferred as the most effective and resilient. RotterZwam reached a financial break-even in 2015, after a few months of activities. Since then, they have been paying salaries to two directors and, recently, three employees. The next step is to replicate their model of enterprise in other cities rather than expand their Rotterdam farm. Their experience is shared with other farmers, and they have estab-

[16] https://www.youtube.com/playlist?list=PLhXTewbEDkBdl4ZjYRAgaJO8z9rNCH4ve

[17] https://www.youtube.com/watch?v=H8iUvE5g1a8

lished a Mushroom Learning Network[18] that functions as a forum. They run training schemes and last year they had three trainees, totalling forty entrepreneurs to date who have been trained to start their own farms (Table 6.10).

Environmental Performance The summary of the environmental impact report 2018 (the report for 2019 is still in development), that can be viewed on the company's website refers to the Sustainable Development Goals as indicators of success, although the summary does not clarify how these goals are linked to the quantified impact of the farming activities. In detail, in terms of transport, mushrooms are delivered within a radius of 10 km, using electric cars. The new farm deploys 156 PV panels on its roof providing more than their operational energy needs. In summer, for example, the energy generation can reach 120 kWh/day. The farm uses 15–30 L of water for 1 kg of mushrooms. 7822 kg of spent coffee grounds were used to produce 959 kg of mushrooms (data referring to 2017 since there was no production in 2018). The substrate for the cultivation of mushrooms is composed of spent coffee grounds, coffee husks (which are released during the process of roasting coffee beans and regarded as waste) and straws. RotterZwam is looking at ways to compost it or use it as animal fodder. Part of it is composted on site with the aid of worms. RotterZwam have built a network with local roasters to collect coffee husk. Fresh spent coffee grounds are preferred, as these do not require pasteurisation, which is an energy intensive process. The documentation of this performance demonstrates a level of responsibility and transparency, which is uncommon, although not all data is available. For example, the total energy consumption is not reported. Yet, the scrupulous yearly quantification of resource use suggests a sincere intention to use this assessment as a tool for improvement of the overall performance.

Table 6.10 RotterZwam: productivity and environmental performance over 1 year (2019–2020)

Rotterzwam	Commercial mushroom farm
Objective	Commercial
Employees/volunteers	Two directors + three employees
Total size of the farm	650 m^2 + offices
Production units and area	238 m^2
Initial estimated investment of the farm	€ 300,000
Energy consumption	ca.45,000kWh/year
Energy production [solar panels]	ca. 50,000kWh/year
Water consumption	N/A
Mushroom harvest per year	18.000 kg/year
Market value	€ 162,000
Other products and services - quantities	A lot of products and services.
Market value of products	+/– €400,000

[18] https://www.facebook.com/groups/Mushroom.Learning.Network.Discussion.Group/

6.11 Trueffelwerk (Mushroom Farm)

Director: Sarah Kuper
Location: Recklinghauser, Germany
Website: https://www.trueffelwerk-recklinghausen.de/
Visited on: July 2019

Motivations Initially, Sarah Kuper created Trueffelwerk to preserve the right to live on the farm rather than because of an interest in food production. She had just bought a farm in the peri-urban area of Recklinghauser, a city with a population of 115,000, within the densely populated Ruhr region. She moved there with the family to live in closer contact with nature. According to German legislation, farmland must only be occupied by farmers, a law designed to avoid speculation and ensure agricultural land is used as such. Forced to begin a farming activity or leave the farm, she consulted a few experts before deciding to start the cultivation of mushrooms in the garage/barn of the farm, in 2018. The mushroom farm occupies two rooms (Fig. 6.22), with the addition of a selling area located at the entrance of the garage/barn (Fig. 6.23). She trialled and is still experimenting with ways to improve her production technique but since the start of this activity, financial viability was the primary criterion driving her choices. For example, the oyster mushroom, which is the only mushroom produced at the time of the visit, was chosen because of its high value on the market.

The entire cycle of mushroom farming is organised inhouse, including spawning. Sarah keeps a culture of spores to germinate in a mixture of rye and wood shavings. When the mixture is colonised, she transfers it to a straw medium, mixed with a cement mixer. She subsequently fills long and narrow plastic bags, hanging them on

Fig. 6.22 Trueffelwerk growing room

Fig. 6.23 Trueffelwerk shop

racks and only when ready, these are moved into the second room that is equipped with a device for mechanical ventilation (a simple air extractor with an electrical fan) and one for humidifying the air to a level of 95% humidity. The equipment of the room includes a basic digital thermometer capable of monitoring the degree of moisture in the air. Sarah also uses big plastic baskets filled with substrate, that sit on the floor of the growing room. The plastic bags hang from metal racks.

Sarah is well informed about the urban agriculture debate, which she unconditionally embraces. She believes she would not have accepted the challenge of starting a mushroom farm, had she not had parents who were concerned about the environment and transferred these values to her. Generally, she believes that nature and urban food growing are new values that are replacing traditional ones such as religious values that are disappearing and leaving a gap. Sarah believes that German society can be conservative and oppose modifications of traditional values, lifestyles and practices, including urban agriculture. But change is necessary to build resilience.

Outcomes Trueffelwerk is self-funded, and Sarah is the only person working in this mushroom farm. She sells her products through her shop, and also participates in farmers' markets. In an attempt to make her business financially viable, Sarah is now farming quails and selling eggs, and has two trees that she is using to grow and harvest truffles. She has designed a range of products such as dried mushrooms, packaged in elegant paper bags. She also produces a powder from dried mushrooms as a flavour-enhancer, which she sells in labelled jars. Mushrooms and products are sold in the shop fitted in the barn, with furniture made from recycled timber. Data on productivity and resource use are not recorded.

Table 6.11 Trueffelwerk: productivity and environmental performance over 1 year (2018–2019)

Trueffelwerk	Commercial mushroom farm
Objective	Commercial
Employees/volunteers	one director + three employees
Total size of the farm	$300 m^2$ + $150 m^2$ shop (estimated)
Production units and area	$100 m^2$ (estimated)
Initial estimated investment of the farm	€100,000 (estimated)
Energy consumption	€5000 per year (estimated)
Water consumption	€300 per year (estimated)
Mushroom harvest per year	3000 kg (estimated)
Market value	N/A
Other products and services – quantities	Jams; dried mushrooms; mushroom powder, quail eggs; cooking classes for mushrooms; participation in wine tastings

In 2019, her business reached break-even. Since then, Trueffelwerk has expanded and has recently employed three women. Produce from the farm are sold, including raspberries, kiwi, quince and pumpkins. Processed food such as jams, honey and E]ierlikör (a traditional German liqueur from eggs) expand the range of products. Products are sold following the Marktschwärmer model, a model which was first created in France, based on the organisation of local food supply chains, ensuring a limit of 60 km from farm to fork (Food Assembly, 2019), The project has had good media coverage. Sarah participated in television programmes such as Lecker an Bord, Land und lecker and Lokalzeit; and was interviewed by several newspapers and magazines (Table 6.11).

Other projects

6.12 Educational Organisations

6.12.1 Sow the City (Educational Organisation)

Director: Jon Ross
Location: Manchester, UK
Website: https://www.sowthecity.org
Visited on: October 2019

Motivations The social enterprise Sow the City was founded in 2009 to promote sustainable food and food growing to schools, mainly organising food gardens on their premises. Over 10 years of activity, Sow the City developed about 150 projects. Recently, they started collaborating with the UK National Health Service by providing consultancy on gardens within sheltered accommodation (i.e., accommodation for elderly or disabled people). Currently, they are retrofitting a small building which will host workshops and community spaces, together with a rooftop greenhouse with an array of raised beds and a hydroponic farm.

One of their objectives is experimenting with food production technologies such as soil-less technologies. In 2016, as part of the events organised by Manchester – European City of Science, Sow the City were asked to design an installation, which they called *The Allotment of the Future*. It was a self-sufficient outdoor hydroponic unit, powered with PV panels. The unit was composed of a metal frame supporting 8 tubes, each 1.2 m long and with 6 plants, with a total of 48 plants. Next to the metal frame was a storage space with the water tank and the PV panels on top. The unit was developed as a demonstrator to show that hydroponics was affordable and could be easily self-build: a way to democratise food technology. The installation attracted the interest of many people, particularly young males, rather than women. Sow the City believe that the technology component is particularly attractive to this age group.

Outcomes The hydroponic unit was not meant to demonstrate high productivity levels, but it helped model productivity and economic feasibility in a study in which Sow the City partnered with the University of Salford (Hardman et al., n.d.). The study, based on a hypothetical 1000m^2 farm, generated a tool identifying capital and running costs for different options of farm. It showed that typically, 8 years are needed to break even and that electricity costs for LED are the biggest cumulative expense over a period of 15 years (£30,000–40,000 per year), followed by labour and rent. It is a simplified tool, allowing – for example – the evaluation of revenues using one crop, rather than a mix of crops. Yet it provides an initial evaluation and identification of options (e.g., PV to reduce energy use) which may impact the business plan. The hydroponic unit of The Allotment of the Future raised interest and was ultimately purchased by the management of the Printworks, a leisure centre in central Manchester with a community garden on its rooftop The facility manager of the company manages the garden and the hydroponic unit with a few volunteers. In 3 years, it has experimented with several crops, including tomatoes and peppers. Harvest is donated to charities.

6.12.2 *Hemmaodlat (Educational Organisation)*

Location: Malmo, Sweden
Website: www.hemmaodlat.se
Visited on: May 2018

Motivations Hemmaodlat ('home grown') is an association founded in 2015 in Malmo – the third most populated city in Sweden. The association aims at promoting and facilitating the use of hydroponics, enabling people to grow their food indoors, particularly in cities where land to grow food is scarce. In a Northern

Fig. 6.24 Entrance to Hemmaodlat from Augustenborg square

Europe climate with long winters, growing crops is limited to a few months over the year. In this context, space-efficiency and access to food all-year-round can be attained only through indoor hydroponics. Before starting Hemmaodlat, the director experimented with hydroponics at home for years, building his own equipment and studying soil-less techniques. With a grant from the municipality, the director developed an educational project based on hydroponics. A commercial space was rented, on the ground floor of a mixed-use building in a square in Augustenborg a low-income, high-density neighbourhood in Malmo (Fig. 6.24). The commercial space would showcase and promote different hydroponic techniques to the Augustenborg's inhabitants, who would particularly benefit from healthy food grown in their flats and homes. Situated next to other shops and fast-food restaurants, Hemmaodlat displays Zipgrow towers in the shop windows, with an impressive array of crops such as onions and runner beans (Fig. 6.25). Since Hemmaodlat activities are solely educational, crops grown are not chosen on the basis of market value but rather as a demonstration of the possibilities that soil-less techniques offer.

Outcomes Today, Hemmaodlat is run as an association that provides training courses and other educational services. Its members (70 at the time of the interview) provide support and expertise not only for these activities but also for experimenting with new and advancing existing soil-less techniques, including aquaponics, aeroponics and anthroponics, which uses urine as a nutrient. Combinations of LED light colours are tested to improve plant growth. Irrigation techniques are tested too and although each system is not completely digitally operated, sensors detecting temperature, pH and other factors are deployed and connected with the smart phones of the directors. This makes Hemmaodlat an innovation laboratory in the field of food technology. The socio-cultural profile of the members varies, although many are

Fig. 6.25 Zipgrow towers in Hemmaodlat shop windows

young and educated (mainly with degrees in engineering or completing their studies in this area), with a keen interest in technology. Media coverage is good, and this has resulted in consultancy projects. The directors are also planning to expand the offer of courses as well as lectures/workshops. For example, they are planning to start a project on food waste.

Chapter 7
The Web Community of Soil-Less Farmers: A Case Study

Valentina Manente and Silvio Caputo

Abstract This chapter investigates the scale of interest that soil-less technologies attract on the internet. While a search for existing urban soil-less community-led projects and small enterprises suggests that these are still rather limited in number, on the web soil-less technologies attract the general public in large numbers, for varied reasons. The web is used here as the space where numbers, motivations and profile of the users searching for information can be gathered. We assume that generally these users are predominantly not professional practitioners looking for specialist advice, but rather individuals scanning the web for new knowledge and opinions. In short, the vast majority of visitors are laypeople wishing to implement small-scale soil-less systems for self-supply, or to start a small activity or simply searching the web for mere curiosity. Search terms used by web users are identified and subsequently used to search on YouTube videos and details of YouTubers. This leads to the identification of the profiles of these YouTubers and the drivers of their interest in soil-less technologies. Findings suggest that practical motivations such as the possibility for an all year-round crop supply and concerns about the quality of food from industrial agriculture attract the largest share of public, while commercial motivations are minor.

7.1 Introduction

This chapter presents an additional case study, in which we identify the scale of interest in soil-less technologies from the general public, using the web as the platform through which people gather information and search for solutions on topics of interest. In this book, case studies and the presentation of the broader soil-less context in Europe suggest that the phenomenon is still small. But interest from the general public is greater than this. The web is used here as the space where numbers, motivations and profile of the users searching for information can be gathered. We assume that generally these users are predominantly not professional practitioners (e.g., SMEs in the food sector) who would typically seek professional advice, but rather individuals scanning the web for new knowledge and opinions. In other words, laypeople wishing to implement small-scale soil-less systems for

The original version of this chapter was revised: the Author "Valentina Manente" has been included now. The correction to this chapter is available at https://doi.org/10.1007/978-3-030-99962-9_10

S. Caputo, *Small Scale Soil-less Urban Agriculture in Europe*, Urban Agriculture, https://doi.org/10.1007/978-3-030-99962-9_7

self-supply, or to start a small activity or simply searching the web for mere curiosity. Therefore we ask: why do laypeople with little knowledge of soil-less technologies decide to experiment with them rather than resorting to conventional horticulture, which can be practiced with simpler techniques and tools? How do people appropriate and interpret technology with regards to soil-less technologies? Do people perceive soil-less technologies as high-tech food production? Do they perceive soil-less as more environmentally efficient than conventional horticulture?

The internet is a highly sophisticated and complex product of technology, designed to be used also by people with limited technical knowledge in this field. It is one of the most evident, contemporary manifestations of the interaction between social and technological systems, which has deeply changed patterns and modalities of social relations by offering the opportunity to form communities of interest that are not limited by geographical boundaries. Searching within the soil-less agriculture community of interest and understanding how individuals portray themselves when they promote their soil-less projects on, for example, YouTube, can provide clues on the motivations behind their interest. These videos enable the identification of the socio-cultural profiles of their producers as well as their personal histories, similar to the case studies presented in Chap. 6.

Google and YouTube are two of the most used repositories of material available on the web. Many of the websites visited and videos viewed in this study are commercial or produced by individuals documenting their achievements in building and running soil-less units. Information provided through these videos is far from being scientifically tested and reliable. Yet it is indicative of a growing interest, an overall willingness to share new knowledge either open access or for promoting expertise and equipment that can be purchased online. Either way, knowledge is produced and made available outside scientific institutions, often by non-experts; a form of democratisation and bottom-up production of 'unorthodox' knowledge, similar to the concepts of alternative technology and frugal innovation outlined in Chap. 3.

7.2 Methodology and Results

This study is mainly qualitative, with quantification of viewers and frequencies provided, and focusing on hydroponics and aquaponics only. Terms that are most frequently searched on the internet were identified and, using these terms as keywords, videos posted on YouTube were selected. The most frequent terms found on the internet were analysed in term of numbers (i.e., how many viewers searched on the particular term) and plausible reasons for the search. For example, a high number of people searching 'home hydroponics' suggests that there is a predominance of individuals who are willing to experiment with this technique at home, either driven by the objective of achieving some form of self-sufficiency or amateur interest in alternative food technology generally. Likewise, the analysis of the videos posted on YouTube enables the identification of some categories under which these videos can

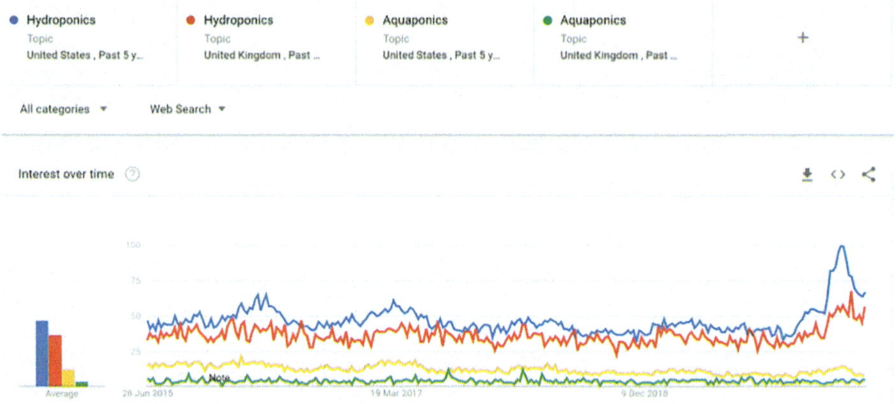

Fig. 7.1 Chart comparing search trends for hydroponics and aquaponics in the UK and the US between June 2015 and June 2020

be clustered, according to the particular interest and motivation or by socio-cultural profiles. The study is limited to the UK and the USA in its Google search, but with no geographical boundaries on YouTube. Samples are relatively small and results cannot be generalised. Hence it must be considered as a pilot aimed at identifying possible trends. What follows is a step-by-step description of the development of the study.

Step 1: Trend search – The keywords *Aquaponics* and *Hydroponics* were used to identify the number of internet users searching for information on these two technologies between 2015 and 2020. Google Trends was used (2020), a search trends feature showing how frequently a given search term is entered into Google's search engine, relative to the site's total search volume over a given period of time.

As Fig. 7.1 shows, over the last 5 years (June 2015 to June 2020), the keyword *Hydroponics* was the most popular both in the UK and the US, with a significant growth in the search rate during the first half of 2020, probably linked to the COVID-19 pandemic. The keyword *Aquaponics* has maintained a constant search rate, probably because more complex as a technology to the majority of web users.

Numbers in the chart (Fig. 7.1) represent search interest relative to the highest point, for a given region and time (Google Trends, 2020b). Each data point was divided by the total searches of the geography and time range it referred to, in order to compare relative popularity (otherwise, places with the most search volume would always be ranked highest); the resulting numbers are in a range between 0 and 100. Different regions that show the same search interest for a term do not always have the same total search volumes.

Figure 7.2 shows the spatial distribution of interest in the USA. Generally, the *Hydroponics* search was higher than the aquaponics one, except in Hawaii, where

Fig. 7.2 Spatial distribution of internet users in the UK and the US

many food security programmes with aquaponics were organised between 2010 and 2016 (Beebe et al., 2020). Search rate with *Hydroponics* was particularly high in the East coast countries and also in Alaska. Search rate with *Aquaponics* was relatively high in countries such as Idaho, South Dakota and Wisconsin, hence not concentrated in any particular broad region. In the UK, *Hydroponics* showed a very high search rate – and *Aquaponics* a very low one – evenly across the country. It is possible that UK internet users perceive aquaponics, in particular fish farming, as far more complex than hydroponics and is consequently less pursued.

Step 2: Word search – The second step of the study was to identify the most researched topics connected to the keywords *Hydroponics* and *Aquaponics*, which can help identify the motivations driving people to search for information. A search engine optimisation software was employed (Ubersuggest) to generate a list of the 100 most searched words on the web over the month of June 2020, which were associated with the keywords. A list of 400 words, 100 for each keyword and location, was produced and subsequently classified as shown in Table 7.1. In the UK, the search volume of a single month was 53,720 (45,780 searches for hydroponics and 7,940 for aquaponics); in the US this figure was 351,810 searches (269,120 searches for hydroponics and 82,690 for aquaponics). These numbers do not necessarily correspond to the actual number of internet users: a user may have searched more than once with different words or the same one. Yet, they provide a broad scale of the interest raised by these technologies. The two countries show a different ratio hydroponics/aquaponics. In both cases, aquaponics is searched for less as a soil-less technology than hydroponics. However, the ratio for the UK is 5.7:1 and in the US is 3.2:1, thus suggesting a higher popularity of aquaponics in the US.

Table 6.11 shows that specific information on components and system design made up for the largest share of searches (50% – Components, Construction and Modifications), followed by learning the general principles of soil-less systems

Table 7.1 Categories and subcategories of words searched on the internet, under the two key words Hydroponics and Aquaponics

Categories	Subcategories	UK Aquaponics	US Aquaponics	UK Hydroponics	US Hydroponics
Components	Spare parts	10	12	8	9
	Automatization	1	0	0	0
	Grow media	2	3	4	3
	Plants and fish	17	16	7	7
	Nutrients	1	2	4	4
Construction	Construction of the system	10	8	9	8
	Type of system (e.g., ebb and flow)	4	6	5	8
Modifications	Hacks (i.e., customisation of existing systems/ types)	9	10	4	9
	Troubleshooting	1	0	1	0
Commercial	Purchase	9	10	18	20
	Job search	1	4	2	0
Knowledge	Literature	11	11	23	25
	Courses	11	8	2	2
	Other questions (e.g., *hydroponics in London* or *aquaponics 2019*)	13	10	13	5

(34% – Knowledge) and locations where components can be bought or job availability (16% – Commercial). This suggests that about one third of the users, who searched for information about soil-less systems has little prior knowledge on this subject area (probably approaching the subject for the first time), whereas 66% of internet users were already at a more advanced stage of knowledge.

The search engine optimisation software can generate a list of the 10 most visited websites connected with the search words. In the UK and the USA, such websites were mainly commercial companies and, when featuring a tutorial section, these were connected to videos that were available on YouTube (see Sect. 6.5 – Mangrovia Scicli and Sect. 6.10 – RotterZwam). YouTube videos represent an effective low-cost tool as they are relatively easy to create and organise.

Step 3: YouTube search - The list of 200 words per country obtained in Step 2 was used to search on YouTube. Videos selected were the most viewed under each word, even if – for some words – they had a low number of views. Some searches led to the same video. Also, videos of organisations were discarded whenever exclusively selling materials, components or courses. Only videos presenting existing operating soil-less units were included. Videos thus shortlisted were 165.

Step 4: Identifying profiles of YouTubers – Profile of YouTubers were determined through the features that some YouTubers shared – e.g., social background, motivations and interest in soil-less technology. Geographic location and qualitative information were obtained through the location feature and the "About" section in the YouTubers' channels, while other data was inferred from the videos' setups, the YouTubers' statements, and other details on their social media pages (Facebook and Instagram).

The identification of six types of soil-less YouTubers helped understand their socio-cultural background. The sample included 142 videos only, as 23 did not provide sufficient information to identify as a type. These types and profiles are as follows:

Self-sufficiency advocate (**SA** – n = 32), generally living in suburban areas or in isolated places, using soil-less techniques for self-supply complementing low income or in line with a lifestyle choice;
Hobbyist (**H** – n = 39), motivated by curiosity and interested in soil-less technology among other activities. This is especially true for hydroponics, which is simpler to assemble and manage;
Farmer (**F**– n = 5), documenting their practices and advertising their business. Soil-less techniques are seen as complementing traditional growing methods. Motivations mentioned vary from the desire of going back to a more traditional lifestyle to voicing environmental concerns;
Suppliers (**S** – n = 21), marketing products or services;
Off-gridders (**O** – n = 7), living off-grid either for economic reasons or lifestyle choice; and
Educators (**E** – n = 38), promoting training courses, seminars and manuals.

Within this sample, the majority of YouTubers were Hobbyists and Educators (27% each), followed by Self-Sufficiency advocates (23%). Suppliers had a moderate presence (15%) while Farmers and Off-gridders represented only a small share. The sample of aquaponic YouTubers showed a higher share of Educators (36%) and a significantly smaller share of Hobbyists (22%). Conversely the sample of hydroponic YouTubers attested a prevalence of Hobbyists (33%) followed by Self Sufficiency advocates (26%) and Supply resellers (18%), while the number of Educators considerably shrunk in this sample (17%). In our sample, aquaponic YouTubers seemed to be expert in this area and utilised their units as demonstrators rather than food production and supply (Fig. 7.3).

Step 5: Interrogation of the profiles. A smaller sample of videos was selected in order to further analyse motivations and approaches to soil-less technologies. YouTubers selected were providing sufficient details about the aim of the project, their background, the motivations and the context driving the project, and views on soil-less technologies generally. The final sample totalled 30 YouTubers: 17 hydroponic farmers and 13 aquaponic farmers. 25 additional videos were reviewed, which were linked to the YouTubers' videos sampled and provided further relevant information.

In this final sample, most of the profiled YouTubers were white, with a small percentage of Asian and Black speakers. This could be ascribed to the fact that this research was conducted employing English keywords that may attract mainly

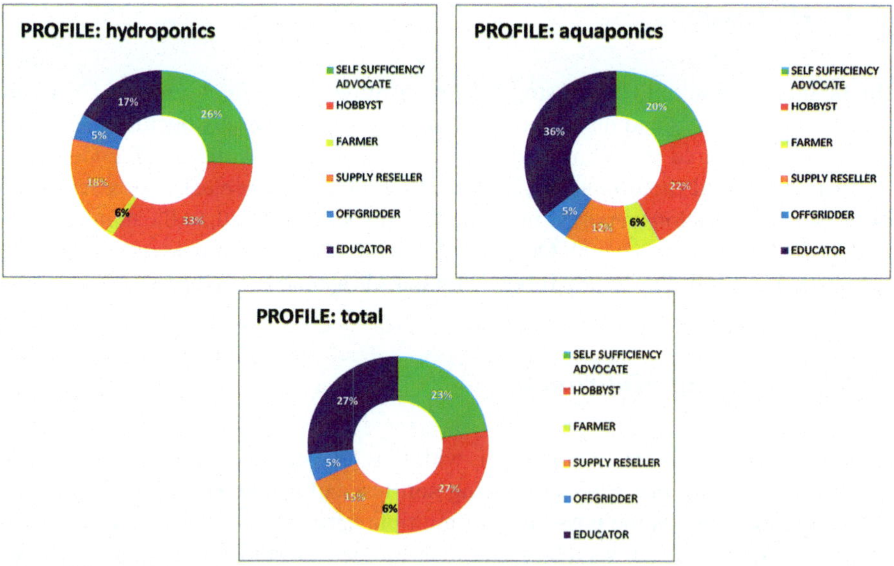

Fig. 7.3 Profiling of the 142 videos

people from English speaking countries with a predominantly white population (Dos-Santos, 2018). Most of the speakers were based in the US, both for hydroponic and aquaponic projects, while Australian speakers were more frequent in videos showcasing aquaponic projects, with only one showing a hydroponic project. Canadian YouTubers were only presenting aquaponic videos, and Asians represented a small number for both technologies. Finally, the vast majority of YouTubers were men, with only 2 women out of 30. This seems to contradict a survey among urban farmers in Maryland, USA, in which 52% participants were males and 48% women (Little et al., 2019). In rural farms, studies suggest a strong gender imbalance: over the last two decades, in US family-led rural farms, the role of women has been secondary and invisible, even when they actively contributed to farming (Fremstad & Paul, 2020; Leckie, 1996).

Aquaponic YouTubers were mostly in their 30s and 40s, but those advocating hydroponic systems were rather more evenly distributed across all age groups. Studies on the demographics of conventional urban farmers in the Global North suggest that the majority are of an older generation. For example, in Japan, where urban farmers account for 25% of farming households nationally, the age of farmers is rapidly rising (Moreno-Peñaranda, 2011). In Bonn, the average age of respondents to a survey of conventional urban farmers was over 50 (Hirsch et al., 2016); the average age of a sample interviewed in Milan was 66, and the majority within the sample (87%) were retired (Ruggeri et al., 2016). Similar conclusions, with an average age of 56 years, were also reported by a study on urban agriculture conducted between Ljubljana, Milan and London (Glavan et al., 2018). The average age of our sample of YouTubers seem to suggest that soil-less technologies can attract younger generations.

7.3 Discussion

In the final section of this case study, the motivations of the final sample of YouTube videos are unpacked and discussed, working towards an overall conclusion. The discussion is structured according to the categories summarised in Table 7.2. Generally, farmers expressed more than one motivation behind their projects. As a result, the total of the motivations reported in the table is higher (n = 77) than the number of the videos included in the sample. Quotes from videos are reported below to support the analysis. The list of the 30 YouTubers with links is available in Appendix B.

7.3.1 *Practical Motivations*

Practical motivations are predominant, with 63% of YouTubers mentioning them as the most important factor for embracing soil-less cultivation (Fig. 7.4a). The most frequent motivation is the possibility of multiple harvests over the year with a share of 26%. "Aquaponics grows a lot faster than soil beds, just like hydroponics. There's a reason people grow commercially in aquaponics and hydroponics and that's because you can turn over the plant a lot faster", maintains YouTuber 19 (video with 76,470 views). This motivation can apply to cultivation in greenhouses generally, whether or not plants are grown in soil. Yet, the decision to use soil-less technologies must be driven by the perception that these are more productive than indoors greenhouse cultivation. The main productive advantage that aquaponics offers, the harvesting of animal and plant-based food within the same productive cycle, is not mentioned, perhaps because many of these YouTubers do not grow food for subsistence and are therefore not interested in the potential of aquaponics to generate within one system food for a complete diet. YouTuber 8 is an exception, observing that: "we started building our second off-grid property around May 2016 and we are loving every challenge of setting up everything from scratch. We are very self-reliant and everything is done with little to no outside help." This video was viewed 220,353 times.

Table 7.2 Motivations of the YouTubers grouped by profile

Motivations		SA	H	F	S	O	E	TOT
Practical	Multiple harvests	2	4	0	1	1	2	10
	Low cost	1	2	0	0	2	0	5
	Ease of maintenance/ease of cultivation	3	4	0	1	2	0	10
	Space constraints	4	0	0	1	0	0	5
	Climate unsuitable to traditional methods	0	1	0	1	3	1	6
	Integration with conventional agriculture	0	0	2	0	0	0	2
	All year round availability	2	4	0	1	3	0	10
Quality	Safety of produce	4	2	0	1	0	0	7
	Quality of produce	2	2	1	1	0	2	8
	Environmental concerns/Social concerns	1	1	0	0	0	1	3
Technology	Desire to experiment	0	3	0	2	0	0	5
Commercial	Production for trading purposes	0	0	0	3	0	3	6

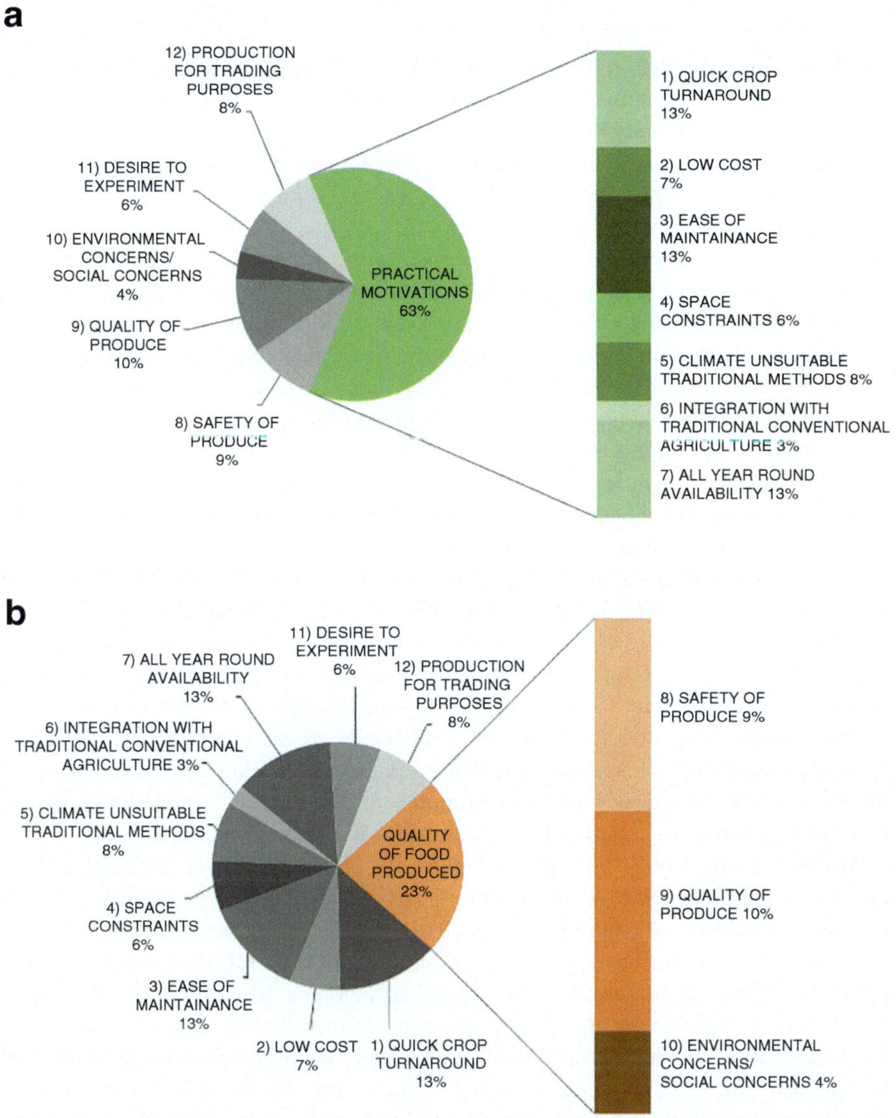

Fig. 7.4 Pie-chart with shares of practical motivations (**a**) and quality of food produced (**b**)

The prevalence of productivity in this sample of YouTubers goes against the spirit of many of the community-led soil-less projects presented in Chap. 6, which generally embrace a strong social agenda. This can be related to the fact that videos sampled present projects organised by individuals rather than groups. Yet, urban agriculture practiced at an individual level has a strong social component too; allotment sites are places for socialising and community building (Cattivelli, 2020). This is not only a consequence of the gardening practice but also of particular spatial arrangements. Sharing the same green place with other gardeners, inevitably leads to connecting with them. And although at a community level a greenhouse with a soil-less unit can be shared by many, it is difficult to imagine how a soil-less arrangement similar to allotment sites (that is, the aggregation of many plots where food can be grown individually) can happen. Perhaps a conclusion from these reflections is that, in order to generate social benefits, soil-less food growing at an individual level requires new approaches for sharing space and equipment.

Ease of maintenance is mentioned by 13% of the sample. YouTuber 29 says: "I don't need to worry about watering plants, I don't need to worry about fertilizing them and everything" (channel with 37,900 subscribers). This view shows how soil-less techniques lead to a very different approach to horticulture, in which plants are not managed with conventional irrigation methods, perceived as 'worries' by this YouTuber. Yet, soil-less farmers will inevitably need to dedicate time to maintaining the equipment, rather than the plants. Affordability and Space Efficiency are motivations mentioned by shares of 7% and 6% respectively. YouTuber 27 (379,249 views) says: "aquaponics (…) uses the same water again and again, (…), so even with the water restrictions that we have here in the valley where I live (…) it's not even a blip on our water bill and that's pretty amazing". YouTuber 13 (22,284 views) turned to soil-less techniques after relocating to the suburbs, where farmers are faced with several space constraints "I had to sell my 17-acre, rural homestead and return to the suburbs". YouTuber 19 (91,473 views) says that soil-less techniques are the only way to attain food self-sufficiency for some farmers: "I've got a friend in Florida that lives in a community (…), he can't raise animals on his suburban block (…) but he can do aquaponics (…). Sometimes your hands are tied and you just got to go with what works and for some people that's aquaponics ". Space-efficiency is one of the greatest advantages of soil-less techniques and these two YouTubers recognise that this is an advantage particularly relevant for urban environments.

Another motivation, which is relevant to 8% of YouTubers, is the local climate and the possibility to save water in water-scarce contexts such as Australia. A case in point is YouTuber 20 stating on his website "If you live in drought affected areas (as much of Australia has experienced over the last decade), then I believe there could be some significant benefits using aquaponics to reduce water usage". YouTuber 21 explains his aquaponics systems are installed on a farm as one of the food cultivation techniques available, which can complement more conventional techniques, one not excluding the other. This is true for the 3% of the surveyed YouTubers sample. It is a view of an integrated approach to agriculture in which technology and conventional methods go hand in hand ("We've built an aquaponic system on the farm inside a 600 square foot greenhouse, recently converted to a hydroponic system."- video with 44,789 views). Other practical reasons are connected to the reduced physical effort. An Australian Off-gridder (YouTuber 8 – 220,353 views) presenting his hydroponic system says: "we're too old (referring to the act of digging and planting). No muscles involved". Although expressed by a small share, this motivation suggests that community-led soil-less techniques are fit for older urban farmers, indicating a direction of development for these techniques that is novel and relevant to a large group of urban dwellers.

Overall, the videos show a hands-on approach to technology. Most of the YouTubers combine hardware store materials with specialised tools and often document their trial-and-error approach on a personal channel while experimenting with new techniques. Users that cultivate with hydroponic technologies are creative and eager to experiment with components and layouts. For example, YouTuber 11 (8489 views) shows how containers such as coffee jars were used for planting. YouTuber 4 explains how their approach to the construction on the replication and modification of other systems: "...this is the system I built, I saw some other systems online and I adapted what I saw into this..." (2166 views).

Aquaponics farmers tend to opt for a more conservative approach, possibly for the complexity of this system which requires a considerable initial monetary investment. Moreover, fish safety which could be threatened may also deter from experimenting further. YouTuber 19 (76,470 views) remarks "I made a very simple and silly mistake that cost the lives of 10 fish that have been in the system from the start".

7.3.2 *Quality of Food Produced*

The quality of food produced is a motivation for a smaller share of YouTubers. 19% are concerned with the overall quality of the food they consume generally and utilise soil-less technologies since these enable full control of the whole production process, as remarked by YouTuber 19 (75,817 views): "we're growing fish to feed ourselves and we're growing fish we know where it has come from, we know there's been no (...) antibiotics or other chemicals added in there. That happens in commercial farmed fish; so that's one of the big bonuses." In spite of the traditional mistrust towards input intensive industrial agriculture or even non-conventional food production methods (eg. GMOs) (Poortinga & Pidgeon, 2007; McWilliams, 2014), the use of synthetic nutrients for crops is generally accepted. Yet some YouTubers stressed how the soil-less technologies produce organic food, as remarked by YouTuber 20: "... aquaponics provide food at its maximum, food from plants as well as the fish. (...) No need for artificial fertilizers (...) it's a complete natural and organic system". In this case, the use of the word organic does not refer to the organic standard which is still not applicable to soil-less produce. YouTuber 20 is a commercial enterprise promoting their services through the video. Their claim of organic produce is perhaps knowingly inaccurate and used for marketing purposes. However, YouTuber 24 (1,414,754 views) states that since synthetic nutrients for hydroponics are not "organic", other chemical compounds that are deemed as "natural" can be used to grow 'natural' crops "you'll hear some people say – oh you're supposed to use potassium hydroxide in there – well that's not an organic compound...I use nothing more than over-the-counter vitamins".

YouTuber 2 referred to food security as one of his main drivers to start a YouTube channel on hydroponics (46,400 subscribers). "Food is a basic necessity of life and every single person should not have to worry about their next meal", while YouTuber 5 included resource consumption among the reasons for preferring hydroponics to traditional growing techniques: "I know some people are not fans of this stuff: we got plastic, chemical nutrients, artificial lighting. But you know? It's amazing growing plants in here. They smell great, they clean the air, they look amazing. To me this is the most responsible use of resources." (94,393 views). Only 4% of YouTubers expressed motivations connected with the environment and pollution. It is surprising that sustainability motivations are apparently not popular within this sample of YouTubers. They are certainly at the core of many urban agriculture projects and, once again, soil-less technologies are promoted as resource efficient, which is one of the conditions for sustainability. It is difficult to establish whether the connection between resource efficiency and sustainability is not clear or whether this is so obvious that it is not worth mentioning. But the predominance of practical motivations, together with misrepresentation of terms such as natural and organic, and how these apply to soil-less produce, suggest that the sustainability related implications of soil-less technologies are not completely comprehended.

7.3.3 Relationship with Technology

The category of technology was identified as desire to experiment. This was manifested by quite a small share of the sample (6%), mainly Hobbysts and Suppliers. It is evident that YouTubers within this sample accept soil-less technologies unconditionally. Therefore, evidence of technology as a motivation was identified with the intention of engaging with it to a further extent. In this share of YouTubers, the relationship with technology is mainly displayed through the propensity to experiment with techniques and assembling methods. YouTuber 11, for example, tests different aspects of soil-less technologies by employing household props, and documents her progress in her channel. "I think this experiment shows that the hydroponic method will produce the same or even better results than the soil method" (video with 24,930 views). Another attitude that can be associated with fascination with technology is functionality. YouTuber 24 (61,251 views) reflects: "I'm actually more concerned with functionality: I get so much food out of these things that I honestly don't care what they look like; functionality is the important part". It is quite surprising that profiles such as Farmers and Self-sufficiency Advocates do not express technology as a major motivation. Being profiles concerned with productivity, they are likely to engage with techniques and technologies to enhance functionality. It is possible that this process was not perceived as a driving motivation but rather as a basic component of the farming profession: one which is inevitable rather than aspirational. Yet, the analysis suggest that technology is attractive only for what it can offer and that unpacking such technologies to understand their inner workings is not perceived as a strong driver by the majority of our farmers (Fig. 7.5a).

7.3.4 Commercial Reasons and Other Observations

The share of YouTubers moved by economic motivations is only 8%, although an underlying advantage not directly stated but implied in many videos from other Youtubers is the potential of soil-less technologies to produce fast growing,

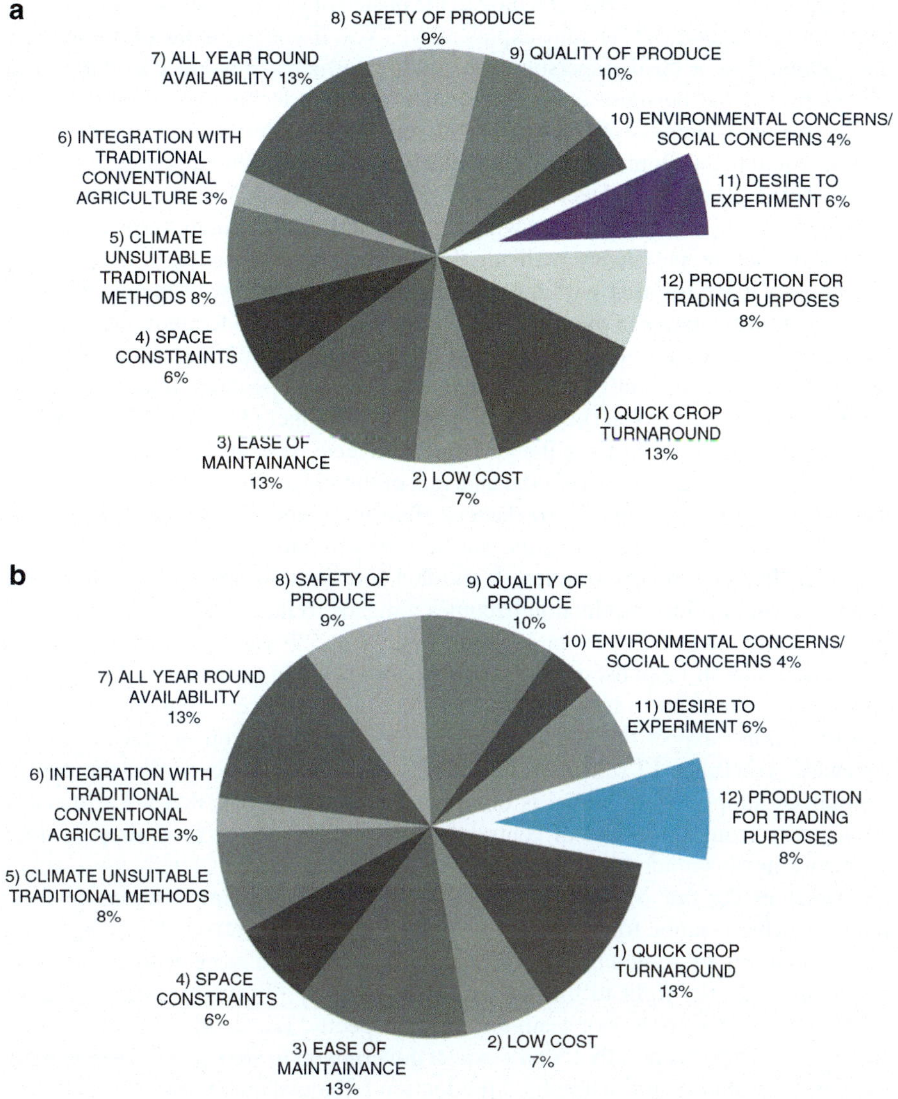

Fig. 7.5 Pie-charts with shares of motivations in relation to attitude towards technology (**a**) and commercial reasons for soil-less techniques (**b**)

high-value crops. Advice given by some YouTubers in this group stresses the importance of a correct economic strategy for the survival of the farm. YouTuber 23 says; "When you're planning a new farm business, you want to consider your production methods and practices from the very start. Production practices shape your business plan, farm construction, and everyday operations for years to come". YouTuber 30 emphasises the importance of marketing strategies: "Making it pay is all about having the product that people want and getting it to them. This is the part of the plan that should occupy 80% of our overall planning, implementation and ongoing enquiry".

The response in the YouTuber 22's blog to the posted question "Can you make money from doing aquaponics?" is quite explanatory: "Yes. If you put in the effort and conduct yourself as a businessperson more than a farmer". Beyond the enthusiastic statements on the business opportunities that soil-less technologies open, caution transpires: an awareness of the fragility of small-scale enterprises operating in this sector, which must adopt innovative approaches to attract clients in order to survive.

YouTuber 30 suggests leveraging on health and food safety issues: "The health and better living movement is growing rapidly, and it takes many turns as it does. Like never before, people are realising that food purity and quality is the pinnacle, the thing to strive for, that will improve and restore personal well-being. We aquaponics practitioners are in the health food business". Soil-less technologies can also be presented as a way of generating income through the home-growing of highly profitable plants that don't take up much space such as peppers, microgreens and tobacco. For instance, YouTuber 16 promotes his channel (410 subscribers) as one advising on how to profit from the growing of diverse types of vegetables.

There are other findings from the analysis of the videos that are not connected to the four categories presented. YouTubers often refer to other people's videos as a way of completing or expanding the information provided. This shows that there is an established community of interest, and that social media are used as a forum. In many videos, soil-less production becomes a means to independence from a social system or the market. For example, YouTuber 9 (63,539 views) launches a call for self-sufficiency in response to recent food shortages due to the COVID-19 crisis: "do whatever it takes to be self-sufficient, look after yourself, look after the ones you love, don't expect the government to step in and help you in times of need". Likewise, YouTuber 17 (channel with 9007 subscribers) believes that the current centralized supply chain model is unreliable: "It's my feeling that we all need to move away from the model of centralized distribution of our food supply which makes us far too vulnerable. It's my opinion we need to be increasingly more reliant on ourselves for our own well-being". There are political implications in these remarks, which can be found in many urban agriculture initiatives. However, while these initiatives are rooted in social movements asking for a more democratic use of resources (e.g., the right to the city and food security (Tornaghi, 2014; Sonnino, 2009), our YouTubers do not seem to focus on specific issues other than to express feelings of distrust generally ("don't expect government to step in"). In between this extreme view and groups using food production for social amelioration, other motivations can be found that rely on soil-less to be off-grid, aspiring to a lifestyle that is detached from an undesirable society. This is clearly represented by YouTuber 8's comment where he declares his independence from external subsistence networks (220,353 views).

One of the findings from this case study is that the interest in soil-less technologies is growing not only for commercial and community-based projects but also among the general public (considering internet users as such). The number of views of the YouTube videos gives an indication of this scale of interest, with some videos reaching over 1 million views, at the time when this study was developed. Within the general public, practical information is the main driver for turning to Google as a repository of knowledge. Issues such as the elements composing soil-less systems, the way these systems are engineered, or simply literature on soil-less technologies, constitute the most frequent searches. Moreover, practical motivations such as ease of cultivation and advantages such as high yield were the priorities behind most of the projects showcased in the YouTube video sample. Such motivations are perhaps not in line with those behind many of the farmers interviewed for the case studies, who generally prioritise motivations related to sustainability and education. However, this discrepancy is also a consequence of the limits of the research design for this web-based case study. The 30 videos selected were those offering a complete picture of existing projects as well as a complete profile of farmers. This criterion is likely to result in a sample that is not representative of the wider YouTube, soil-less farming community. In fact, within the broader sample used to identify soil-less farmers profiles, Educators numbered 38 out of 142 (27%). Self-sufficiency Advocates and Off-gridders, implicitly motivated by concerns about the current rate of exploitation of resources, totalled 39 (28%). Together, they represent most of the videos sampled.

Hobbyists and Suppliers were 43% of the sample. These two profiles are likely to be particularly interested in the technological and engineering aspects of soil-less cultivation. While educators, environmentalists and technologists seem to be well represented, the smaller group fell under the profile of Farmers (5 out of 142). In many of our case studies, the internet, social media and YouTube were important tools for self-promotion and outreach strategies. While no general conclusions can be drawn because of the limits of the sampling and the sample size, the imbalance between the number of professional farmers and the other profiles is striking. Many reasons may inhibit farmers from producing a video and uploading it on to YouTube (e.g., time, business models not requiring social media for marketing strategies and scepticism towards social media). Yet, soil-less farmers were generally younger and likely to be familiar with the internet and information technologies; the small number of YouTubers under this profile may as well be representative of the real share of professional farmers that are producing food with soil-less technologies. Also, the number of Suppliers is quite large (21). This suggests that demand for soil-less units and components comes from varied groups, including practitioners and amateurs, not only professional farmers. It also suggests that commercial practitioners may not be the largest market share in this sector.

Within its limits, this 'virtual' case study completes the picture that the case studies in Chap. 6 give about the characteristics and size of the small-scale soil-less communities in the Global North. It is still an unclear picture in which soil-less technologies are not consolidated and are used for a wide range of purposes. The many profiles identified suggest a multitude of interests, not always coupled with a clarity of views on the real advantages and drawbacks. Yet it is a dynamic scene, likely to evolve soon and as such, requiring analysis and close monitoring.

Chapter 8
Discussion: Analysing the Case Studies and the Wider Phenomenon of Small Scale Soil-Less Urban Agriculture

Abstract This chapter analyses 12 small scale soil-less urban agriculture case studies in Europe, focusing on the motivations of urban farmers to start these soil-less projects; the actual outcomes of such projects; their productivity (fish and crops); and their environmental performance. The analysis is both qualitative and quantitative. Some of the motivations identified are similar to those that drive many other in-soil urban agriculture projects: self-supply, educational objectives and environmental concerns are powerful drivers for many gardeners and farmers worldwide. This analysis shows in what ways soil-less technologies influence and transform such common motivations, which become strongly connected with the potential of soil-less technologies to, for example, connect with and attract young people, and circumvent risks of producing unhealthy food due to soil and water contamination. Measured against benchmarks borrowed from a study on an experimental unit built by Rakocy and other studies, the productivity of these projects is generally low, with few notable exceptions, possibly because of the limited expertise in these technologies of many urban farmers. The small scale soil-less urban agriculture sector is nevertheless promising and the case studies show great dynamism and determination to improve both productivity and resource efficiency. The sector shows resilience in the face of difficulties such as accessing funds (in the case of community-led projects), or competing on the market (in the case of commercial operations), often by diversifying their offer.

8.1 Introduction

This chapter analyses the case studies in Chaps. 6 and 7. This is not an easy task: the case studies differ from each other in scope, operations, objectives and outcomes, making comparisons difficult. The number of case studies is limited, and results from the analysis cannot be generalised. Nonetheless, using the same interpretative grid as the case studies (Motivations – Outcomes – Productivity – Environmental performance), some qualitative and quantitative reflections are formulated. In this chapter, motivations and outcomes are elaborated and, whenever possible, connected to general trends or compared to other studies. When data for environmental

S. Caputo, *Small Scale Soil-less Urban Agriculture in Europe*, Urban Agriculture, https://doi.org/10.1007/978-3-030-99962-9_8

Table 8.1 List of case studies with characteristics and objectives (projects marked Active where operating at the completion of the book, in April 2021)

Name		Status	Type	Objective
Milagro de los Peces	S	Active	Community-led project	Educational – self-supply
Huerto Lazo	S	Active	Community-led project	Educational – Training facility – commercial
Real food	UK	Active	Community-led project	Educational – self-supply
UGH	D	Not known	Community-led project	Educational – community building
BioAqua	UK	Active	Small farm	Commercial – consultancy
Mangrovia Scicli	I	Active	Small farm	Commercial – consultancy
Smart farmers	B	Active	Small farm	Consultancy – research & development
GROWx	NL	Active	Small farm	Commercial
GrowUp urban farms – unit 84	UK	Closed	Medium enterprise	Commercial
Bristol fish project	UK	Closed	Community-led project	Educational – research & development
RotterZwam	NL	Active	Small farm	Commercial – consultancy
Trueffelwerk	D	Active	Small farm	Commercial
SowtheCity	UK	Active	Association	Educational
Hemmaodlat	SW	Active	Association	Educational – training

performance and productivity are available, they are used to speculate on the real advantages and economic viability of these projects, thus providing some clues on the potential success and drawbacks of small scale, soil-less urban food production. Some of the motivations identified are similar to those that drive many other in-soil urban agriculture projects: self-supply, educational objectives and environmental concerns are powerful drivers for many gardeners and farmers worldwide. This analysis shows in what ways soil-less technologies influence and transform such common motivations.

The following sub-sections (Educational, Self-supply and Commercial) correspond to the objectives (and motivations) that were most recurrent within the projects, summarised in Table 8.1. The other sub-sections (Sustainable food production, Environment and Technology), correspond to underlying motivations that were not explicitly expressed but were clearly driving the choice of a soil-less technology to grow food and the impact of such a choice.

8.2 Motivations

Case studies include four distinct types, differing in organisational characteristics: community-led projects; small enterprises; medium enterprises and associations (see Chap. 5). Each type operates following different agendas. There are similarities in motivations between types, although the same motivation is understood from slightly different angles depending on the commercial/non-commercial purpose of the project or other factors. Table 8.1 shows the characteristics and objectives of each case study. What follows is a list of recurrent motivations and how these are understood by diverse actors.

8.2.1 Education

Education is one of the most recurrent motivations; a stated objective for all the community-led projects and associations in these case studies, but also one that is implicit in the activities of some small enterprises. For example, GrowUp Farms scheduled weekly visits in which the farm was open to visitors. Soil-less units are a powerful educational tool, enabling the observation, indoors and in an urban context, of plant and fish development processes at all growth stages. Hydroponics has been previously used in school environments as a pedagogic device. Hershey (1994) reported that this was happening as far back as the beginning of the twentieth century. Junge et al. (2019) documented how aquaponic units are used in primary and secondary education across Europe, claiming that these can enhance teaching in scientific subjects (Science, Technology, Engineering and Mathematics). Soil-less technologies do not only enrich scientific knowledge but also broaden environmental awareness. A study by Burt et al. (2020) suggests that the use of hydroponic units as part of the science educational curriculum has an impact on students' awareness of issues such as climate change. This approach to teaching can be very effective, allowing a deeper understanding of plant and fish cycles and how these are linked to food production. For younger generations, confident users of technological tools, soil-less technologies can be strong attractors. But although effective in supporting teaching of scientific subjects, this educational approach holds conceptual implications that may not be in line with some of the solutions for a sustainable food production and consumption, such as the seasonality of the food consumed and regenerative agriculture, which are strongly promoted for a more sustainable food system and food production (Reisch et al., 2013; Schreefel et al., 2020). Moreover, on closer inspection, out of the school environment, the educational messages of our case studies vary according to the project, underpinning different ways of understanding soil-less technologies.

Indoors soil-less systems enabling all-year-round supply of crops, allow out-of-season produce to be permanently available. But seasonality is still relevant when the environmental impact of food is considered, as food produced in indoor controlled environments (such as tomatoes in the winter months) can be very energy intensive (Theurl et al., 2014). Most of our case studies do not use much energy generally, although for-profit indoor vertical farms with environmental control systems (e.g., GROWx and GrowUp Unit 84) are energy intensive. It is important that, when using soil-less technologies as an educational tool, the potential drawbacks of such technologies are explained. Moreover, in promoting all-year round availability of crops, soil-less technologies may also have broader implications in terms of food culture. Out-of-season crops can be imported, rather than grown in controlled environment greenhouses, and therefore with high food-mileage. Distinguishing between technological approaches (resource-efficient and high energy soil-less) while at the same time promoting local food (whether grown soil-less or in-soil) must be part of the educational objectives. This entails contextualising soil-less technologies within the broader food system and debating their role within it. In this respect, perhaps Real Food (Sect. 6.3) offers an effective educational model, representing soil-less technologies as one of the horticulture techniques. Their garden, which includes in-soil and soil-less cultivation, shows students soil-less technologies as one of the approaches necessary to the attainment of sustainable food systems, not prioritising any of these approaches, but rather suggesting that technology widens the options for urban food growing. Milagro de los Peces (Sect. 6.1) offers another educational message, particularly suited to the local communities connected with this project, underpinning the food security potential that soil-less technologies can provide at a household level, demonstrated through the low-tech, self-built greenhouse, powered with PV panels.

Hemmaodlat's (Sect. 6.12.2) educational model focuses on training, offered in a space which is also used as a research & development lab, testing the effectiveness of different techniques. Their audience is young and specialised, and fascinated by the innovation that soil-less technologies offer. It is a 'scientific' rather than 'behavioural' (Real Food) or 'political' (Milagro de los Peces) approach to education. This way of understanding soil-less technologies can be found in one of the YouTubers in Chap. 7. In his video, YouTuber 16 tested the functionality of an IKEA hydroponic unit (Krydda), identifying drawbacks and assessing the optimal distance between plants, the distance of the plants from light source and the water drainage system. Hemmaodlat shop and laboratory inspire visitors to take a similar approach. Other soil-less projects are not based exclusively on education but have developed an educational model that produces additional financial income. Growing Underground[1] (a hydroponic farm in London) and GrowCycle[2] (a mushroom farm in Devon) offer visits to their farms that can be booked through their websites, for payment of a fee.

For organisations such as Sow the City (Sect. 6.12.1) education is the primary activity. One of their educational projects is to build small units in schools and educate playfully. They implement and train the team that will manage the unit. Other projects such as ALTMARKTgarten, a hydroponic research facility on the rooftop of the job centre in the city centre of Oberhausen (Hortidaily, 2019) have been designed to allow visitors into a dedicated space without directly entering the growing area, in order to avoid the transmission of plant pathogens. Because of its size and location, it is an urban landmark, visible to all passers-by and generating awareness about new food production technologies.

---

[1] https://www.intotheblue.co.uk/experiences/underground-farm-tour/

[2] https://grocycle.com/mushroom-cultivation-courses/

One of the concepts that can be well illustrated and explained with soil-less systems is a no-waste, circular economy approach to food production, based on nutrient recovery. Aquaponics is particularly appropriate as an educational tool for this. Equally, the use of spent coffee grounds as substrate in mushroom farming, which is promoted by RotterZwam (Sect. 6.10) is a process that can effectively convey the circular economy concept. On their websites and through interviews and articles, RotterZwam use their circular economy approach to demonstrate their sustainability credentials and strengthen their market profile.

8.2.2 *Self-Supply*

All the community-led projects documented here have an element of self-supply in their programmes, although generally the educational motivation seems to be a stronger factor. Self-supply is either pursued to access healthy food or to strengthen food security. FAO suggest that aquaponic units are an effective solution to improve food security in the Global South and provides directions to build a household-size unit (Somerville et al., 2014). The contribution of urban agriculture to provide food in times of crises is well known, but, while in the Global South urban agriculture has been consistently pursued to improve food security, in the Global North this objective re-emerged only recently (Opitz et al., 2016), possibly in reaction to the 2007–2008 economic and food crisis (Lohrberg et al., 2016). A 2015 review of qualitative and quantitative studies on food security could not find any specific study focusing on developed countries (Warren et al., 2015). Siegner et al. (2018) reach a similar conclusion although their focus is specifically on community food security. This suggests that food security in the Global North is a phenomenon that relates to access to healthy food for low-income groups, particularly when there is reduced access to fresh food (e.g., food deserts). Hence, not food security in terms of subsistence, but rather 'nutritional' security in terms of healthy diets. Against this backdrop, Milagro de los Peces is an exception and may well represent a growing trend. In a densely built, European city, in a low-income neighbourhood in which public land to grow food is difficult to access, soil-less technology represents an alternative to be deployed in private courtyards.

As noted, this approach has political implications. Milagro de los Peces generates frugal innovation, in that it customises the FAO aquaponic backyard unit for a European context. The innovation that this piece of frugal technology offers lies in the possibility for many to take control of a particular food technology and decide how to utilise it. A simplified piece of aquaponic equipment and a short training course (provided by Verde del Sur) transferring scientific knowledge on plant physiology and aquaculture enable people, rather than experts, to apply technology in order to meet basic needs. In this way, technology is democratised. Some of the YouTubers documented in Chap. 7 (particularly Off-gridders and Self-supply Advocates) turn to soil-less as a technology enabling independent life, especially in areas with adverse environmental conditions such as the Australian outback. For these YouTubers too, technology enables autonomy. Hemmaodlat appeals to a different audience which is likely to be attracted by technology as a means of experimenting with and advancing it, rather than simplifying it. In this case, technology is appropriated by users who are fascinated by the idea of high-tech, healthy food production and keen to apply their knowledge to innovate current techniques. It is no longer frugal technology, but rather one adapted to a more intellectual pursuit.

Technology can bear negative implications too. In the case of food technology, this may limit its uptake for self-supply. There may be ethical concerns about fish welfare, insufficient evidence of the environmental advantages of soil-less (Specht et al., 2019) and concerns for the use of synthetic nutrients in hydroponics (Caputo et al., 2020). But the technology-enabled, self-supply options experimented in these case studies open new possibilities for individual households and community groups in terms of alternative lifestyles and degrees of food independence. Aquaponics in particular can provide a complete diet, including animal protein. Projects aimed at self-supply not only provide some evidence of potential quantities of produce generated but also experiment with ways to integrate self-supply practices within current lifestyles and urban spatial contexts.

8.2.3 Commercial Motivations

Commercial motivations are strong in commercial enterprises and minor in community-led projects. However, even for commercial enterprises, this motivation is always accompanied by environmental and social ones, therefore always infused with ethical values. Enterprises such as BioAqua Farm (Sect. 6.6) started their business driven by the goal of offering good quality food; the approach to indoor food production of GrowUp (6.9.1) and – to an extent – GROWx (Sect. 6.8) – is holistic, making sure that the delivery, packaging and social benefits are all reducing the impact on the environment. Although not exclusively driven by financial returns, these enterprises are determined to compete on a food market that is relatively new and therefore with high risks. Their ethical approach does not prevent the use of marketing tools such as branding and promotion strategies that utilise e-commerce platforms or social media. In order to survive, these enterprises need to diversify their offer, including added value products as mushroom grow boxes or mushroom paste. Their creative approaches generate innovation although this is not always sufficient to ensure economic viability and some fail.

Business models are typically based on growth and the enterprises included in this study are not an exception. With growth comes also the risk of prioritisation of returns against environmental and social objectives, which make these projects so valuable. It is clearly impossible to predict whether with growth and scaled-up production, the business model of these enterprises may change and partially lose their environmental and social objectives. But this makes the case for encouraging small-scale urban farming all the more relevant, similar to the EU policies for rural development and small farms (Peters & Gregory, 2014). One of the characteristics of urban agriculture is that it can use marginal land and expand through urban infill, progressively occupying sites left over from urban development. Urban morphology, and policy and market constraints, influence the size of most of these projects, which remains modest. However, the urban context and the proximity of such projects offer the opportunity to establish dense networks in which members within each network cooperate. This approach may enable the development of business models that can grow by expanding the number of small production units, rather than the scale of single enterprises. Such a model requires the recognition of a market based on cooperation rather than purely competition. Soil-less systems have a big role to play in this model, because they multiply the opportunity to exploit urban spaces by occupying indoor spaces. Whether this model of urban food production can compete with large scale industrial production is difficult to say, but such a model is necessary for the viability of small scale enterprises and community-led projects, which otherwise will be bound either to conform to market logics of growth or to be marginalised by big enterprises.

8.2.4 Sustainable Food Production

Although not clearly stated as a primary objective, this is a motivation shared by all the projects documented in Chap. 6. Like the concept of sustainability, sustainable food production presents multiple facets. Case studies focus on one or more of these facets, with technology playing an important role. Sustainable food production includes issues related to its environmental impact, to food security and to supply and distribution through sustainable food chains. Depending on the type of project, one or more of these issues were motivating and shaping the use of soil-less technology. Generally, enterprises used technology to produce with minimal impact on the environment. For example, aquaponics was viewed as a solution to the overexploitation of fish by Mangrovia Scicli (Sect. 6.5), together with concerns about food safety (fish grown in polluted water). BioAqua Farm and Smart Farmers (Sect. 6.7) used soil-less technology to address respectively quality of produce and resource depletion (freshwater and nutrients) because of food production. Producing food within a non-natural setting becomes for them an approach to reduce the impact of the in-soil industrial agriculture production on the environment as well as the excessive exploitation of marine environments.

Sustainable food production must be understood as such by consumers. With their presence on YouTube and social media, Mangrovia Scicli and BioAqua Farm use information technologies to promote their food and explain their sustainable characteristics to a wide audience. Higher cost of food grown in small farms that utilise sustainable methods of farming can become a barrier to their economic viability. Small farmers rely on quality and sustainability of their produce to increase their competitiveness. This is augmented by creative and flexible solutions for food purchase and delivery. E-platforms and apps are used to organise collection points and flexible delivery options. Although not connected with soil-less technologies, the deployment of information technologies to build a bigger customer base becomes a pre-requisite of competitivity for these soil-less enterprises. The latest pandemic is a case in point. In the UK, demand for vegetable boxes from city farms increased (Schoen & Blythe, 2020), with IT technologies enabling this particular supply model. There is a risk, however, that the sustainable food produced by these enterprises is purchased mainly by medium-to-high income groups, who typically can afford higher prices. In this respect, soil-less food may not be socially sustainable.

Another aspect of sustainable food production relates to the exploration of new theoretical sustainability models such as circular economy. RotterZwam clearly articulate this on their website, providing information on spent coffee grounds as a substrate for mushroom cultivation. Smart Farmers, although not mentioning circular economy as a theoretical framework guiding their activities, are nonetheless experimenting with wastewater, like other projects such as Roof Water Farm (2020) in Berlin. Waste is a resource which is particularly abundant in cities; this makes an experimentation with waste and sustainable food systems greatly relevant to urban agriculture, both to the attainment of low impact, closed loop food production and to the development of business opportunities. Ecology of natural systems is the model inspiring circular economy. In these projects, technology mimics natural processes.

The approach to sustainable food production for community-led projects is predominantly shifting to its social sustainability aspects. Many of these projects see soil-less cultivation as one of the possibilities for sustainable food production, not distinguishing between the several options, but accepting them as equal within a range of possibilities. For projects such as Milagro, Real Food, Huerto Lazo (Sect. 6.2), UGH (Sect. 6.4) and even Sow the City, soli-less technologies become functional to accrue social benefits. This, to an extent, explains the lack of interest in testing the potential to grow higher quantities, with multiple yearly harvests. It also partially explains the relatively low interest in rationalising resource use or even collecting data to document and improve performance. A brief glance at Real Food's reporting shows evidence of the impact of their activities and confirms that the group is more interested in reporting on the number of people reached, meals served through their kitchen and recipes distributed than the quantity of food produced and supplied, let alone the water consumed in producing this food. Milagro de los Peces is a notable exception. The scrupulous recording of data is probably connected to the academic background of the initiator and the support that other academics provide to this project.

8.2.5 Environment

The health of the environment and the role that sustainable food production can play in improving its current state is another motivation (and concern) voiced. Practitioners in the field of urban agriculture are generally well informed about food production and supply chains and their impact on the local and global ecological systems. Some of the practitioners interviewed such as the founders of Bristol Fish Project (Sect. 6.9.2) and the directors of Mangrovia Scicli began to practice after studying and carrying out research in environmental sciences and food systems. They have therefore developed a scientific view on these topics, together with a breadth of knowledge underpinning the scope and purpose of their soil-less projects. For some of the enterprises, the choice of soil-less technologies was based on their resource efficiency characteristics and the low impact on the environment (rather than the advantages these can generate in terms of food production only) and in the face of the commercial challenges connected with such technologies. As noted, community-led projects are driven predominantly by social objectives and perhaps do not show this clarity of vision about soil-less technology as an opportunity to address environmental concerns. One of the assumptions mentioned in the introduction of – and repeated through – this book is that there can be a tension between the objective of enhancing urban ecology and an uptake of soil-less methods. If urban agriculture is viewed as a great contributor to urban ecology and biodiversity, and urban farmers embrace this view, is soil-less a contradictory choice? The lack of a clear stance on this point suggests that no contradiction was perceived; the consequences on the natural environment of soil-less technologies are not fully questioned by the interviewees.

The thesis promoted by Despommier (2010) and embraced by others (Beacham et al., 2019; Kalantari et al., 2018; Al-Chalabi, 2015) is that vertical farming can release pressure on agricultural land, which, when not utilised, can be rewilded with great advantages for biodiversity and the climate generally. For this to happen, an overall vision is necessary, with adequate policies capable of coordinating urban and rural development, national and international food production and supply chains, as well as the entire agricultural and food sector. The awareness of this beneficial potential impact of soil-less technologies on the environment did not surface in any interview. Perhaps the scale of this vision makes it difficult to connect with the day-to-day reality of small scale projects. Or perhaps technology, soil-less technology included, is sometimes accepted with an absence of a sufficiently critical reflection of its consequences. Its application in the 'here and now' distracts practitioners from a deeper reflection on these relations. Yet, this phenomenon is growing; Nature Urbaine has built one of the largest soil-less rooftop cultivations (14,000 m^2) in Paris (Henley 2020). If commercially successful, this farm will pave the way to other enterprises in this sector together with research into the possibilities that soil-less technologies can generate in an urban context.

8.2.6 Technology

Technology is embedded in all the other motivations mentioned above. From being used as a powerful magnet for educational purposes, to an enabler of affordable, self-build food production units, to a provider of means for advanced experimentation with resource use, technology is the common denominator of all the projects documented here. One of the most interesting aspects is the low-budget, self-build character that many of these projects display, which seems to be in line with the idea of frugal innovation and alternative technology noted above. In fact, these technologies, which are typically used in large scale farms, are customised and adapted to fit the small scale and the agenda of each project. In this respect, technology is humanised, showing a closer connection with local social needs, rather than with big challenges with which people may have difficulties relating to on a day-to-day basis. There is an attempt to unpack and understand the inner logic of these technologies in order to make them work in line with the objectives of each group. The advantages of a trial-and-error approach to the appropriation of these technologies are mentioned by Mangrovia Scicli as necessary to a deep, comprehensive process of learning. The inevitable slow pace of this process hampers the velocity of growth of the business, but this is accepted as a positive factor because it allows better informed choices. Similarly, Real Food acknowledges that mistakes were made in running the system and these were instrumental in its improvement. Slow growth and a progressive understanding of technology is also a characteristic of projects such as GrowUp, which, intentionally, started with a temporary prototype to subsequently expand, using the valuable experience developed during the initial stages of their project. The small scale of these projects and their links to community groups or networks of customers is a powerful way of bringing technology closer to people, which in turn allows people to interact and modify the initial technological tool or process.

Some of these projects aim at experimenting with, and advancing, existing technologies (e.g., Smart Farmers and Hemmaodlat), and in doing so acting as a research and development facility. In these cases, experimenting with existing technologies is possible because of the expertise available within such projects, either because of partnerships with experts or because the people leading them have developed considerable expertise in relevant fields. Urban agriculture is now also practiced by qualified people with expert knowledge in scientific disciplines. This is not uncommon in industrial agriculture, a sector that in some European countries is now attracting younger and more educated people with new job opportunities (Cinquemani, 2019), but a rather newer concept in a sector that is not commercially established. This may be the result of an ongoing transformation of the economy in developed countries, in which the instability of the job market and the insecurity of job tenure leads to flexibility in choosing career paths. School curricula cover sustainability principles and the state of the environment. The new generations of professionals may be equipped with a value system prioritising sustainability, the environment and social cooperation. In fact, the involvement of scientists and, generally, the scientific community in urban agriculture projects is now becoming increasingly widespread. This may because many research projects have been developed with a focus on urban agriculture, and a renewed commitment of the research community to public engagement activities, in response to the latest European policy to promote science in society (see Chap. 3), which can lead to a democratisation of technology and probably a higher uptake of technological solutions, integrated in community projects.

Technology is used also for enhancing outreach and promotion. Online networks and digital platforms for food hubs as those utilised by Magnolia Scicli and BioAqua Farm, are powerful tools, which can be exploited at no or limited cost and that are increasingly critical elements for the success of small enterprises and community led projects. RotterZwam, for example, in response to the pandemic, was able to organise online workshops and ensure an income stream, building on its established presence on YouTube with tutorials and promotional videos. The investigation of YouTubers presented in the previous chapter is also a demonstration of the role that information technology plays – and the scale of this phenomenon – not only in self-promoting suppliers of hydroponic/aquaponic equipment but also in establishing and consolidating communities of interest in which knowledge and experience are shared.

8.3 Outcomes

While the previous section elaborated on the motivations of community groups and entrepreneurs to initiate the projects documented in Chap. 6, here the focus is on their actual achievements, the impact attained and the potential direction for growth. Some of the goals identified above were fully realised whereas others are struggling to be met. The discussion is divided into two sections dedicated to enterprises and community-led projects. There are some points in common between the two types, particularly the emergence of a new profile of urban farmer and the vulnerability and precarious nature of the projects. There are studies suggesting that urban agriculture attracts numerous young and middle-aged farmers (May et al., 2019; Pourias et al., 2015). The case studies suggest that this is all the more true for soil-less projects, which require additional skills and scientific knowledge that young generations often already possess, benefitting from school curricula with scientific subjects at their core. Another point in common is the short life and at the same time the resilience of many of these projects, which is a characteristic shared with many other, in-soil urban agriculture projects (Caputo et al., 2016), but made more critical by the difficulties inherent to managing units of production that require specific technical knowledge and skills and/or succeeding in a competitive market. Two of the projects included in the book have closed and others may well be inactive by the time this book is published. The closure of a project, however, is sometimes followed by a new project, which, building on past experiences, takes new directions.

8.3.1 Small Enterprises

Beyond the specific attainments of each project, which are presented in the case study chapter, a broader and shared outcome is their resilience despite their vulnerability to adverse factors such as the highly competitive food market and lack of funding. In Europe, urban agriculture is not recognised at a policy level as a significant player in the food supply chain. This may not be clearly stated in national policy papers, but it transpires from national and supranational food policies or strategies. The Common Agricultural Policy in Europe does not include urban agriculture in its programme of subsidies and generally national food policies do not recognise urban agriculture as a primary source of food production. There is therefore no official recognition and sufficient incentives that may boost either community-led projects or social and commercial enterprises. Commercially, and in terms of job-creation, urban agriculture is therefore a niche phenomenon and within this, soil-less urban projects and enterprises represent only a fraction, with notable successes and notable failures. Inevitably, the soil-less technology component of these projects increases their vulnerability to, for example, a market with low food prices, competition with high-tech large-scale soil-less companies and diffidence from customers in accepting soil-less produce. While large companies operate remotely from customers, and their produce is sold by big retailers with no indication of the technology used for production, small scale enterprises build their customer networks locally; their production technologies are visible. To survive in such adverse conditions requires creativity, perseverance and technical knowledge. RotterZwam, for example, had to react quickly to Covid-19 and find new channels to sell produce which was previously sold to restaurants. Their presence on social media and their ability to mobilise online networks was key to their success. Many of the projects resort to consultancy as an additional source of income. Yet, projects fail. In the face of the recent opening of the 14,000 m^2 hydroponic farm in Paris (Henley, 2020) another one closed soon after its start. In 2018, 2 months after its completion, the 1200 m^2 rooftop greenhouse designed and implemented by Urban Farmers in Den Haag, Netherlands, declared bankruptcy (Hortidaily, 2019).

As noted, policy support is key to the growth of an urban soil-less sector. In fact, the implementation of such an ambitious project in Paris is probably the result of the endorsement and support of the municipality of Paris to urban agriculture generally. Between 2017 and 2018, the program *"Parisculteurs"*, an initiative of Paris municipality in collaboration with local stakeholders, funded 60 projects aimed at greening specific urban sites, many of which were designed to grow food also for commercial purposes. Across the entire region of Paris, the Green Plan (*Plan Vert*, the regional urban greening program launched in 2017 by the Ile-de-France region) can be used to fund gardens and new forms of urban farms such as kitchen gardens, orchards, market gardens, and urban beekeeping (Reynolds, 2014).

Paris is still an exception, and many other European cities and countries are not so committed to promoting this sector. Yet, the obstacles for urban agriculture projects – especially for those necessitating higher initial investments such as soil-less commercial enterprises – do not discourage entrepreneurs and social innovators. The table in Appendix A includes 71 projects and farms, the majority of which are commercial, representing only a share of those operating in Europe.

A further outcome is the contribution that soil-less technologies provide in catalysing the interest of urban farmers/entrepreneurs, with a socio-cultural and demographic profile that is different from the typical in-soil urban farmer. This can attract the interest of diverse groups to the sector. That said, there are issues that need to be considered. The survival of small-scale soil-less enterprises depends on the multiple skills of these entrepreneurs, with mechanical engineering, horticultural, marketing and more becoming necessary requirements for success. When sales of produce do not generate sufficient return, added value products must be included. Trueffelwerk and RotterZwam are cases in point. The reliance on social media and models of e-commerce (Mangrovia Scicli and RotterZwam) are also another emergent characteristic of these soil-less farmers. Making use of the online community that they formed, they organise workshops and training sessions with additional income and possibly new small enterprises stemming from these initiatives.

Another important outcome is the connection of these small enterprises with the local context, which, inevitably, is a result of the short supply chains they offer. It is also the result of training activities or strategic partnerships with local authorities such as that developed by GrowUp Farms in London. Inevitably, if successful, enterprises expand, with a risk of shifting their focus onto efficiency, productivity and larger supply networks. Retaining a strong social sustainability agenda while expanding, as noted, can be difficult. Besides, it is not clear how size impacts the economic viability of urban soilless small enterprises. The director of BioAqua Farm seems to believe that medium size soil-less urban enterprises are not economically viable. Although this point of view is debatable, there are different models of growth that are being tested. For example, GROWx is attempting to implement a network of small vertical farms, close to the locations of their clients across the Netherlands, rather than increasing the size of the existing farm. This may help retain a close relationship with local clients and reduce food miles. RotterZwam, Mangrovia Scicli and BioAqua Farm demonstrate that the idea of locality is evolving. In cities, social media and virtual networks enable the connection with a community of interest with blurred geographical boundaries. In small cities, information technologies facilitate direct contact with producers which would be otherwise invisible, as is the case for food sold by the big retailers. New supply chain and business models are necessary to preserve the social advantages that small scale enterprises can generate locally. A debate on scale is necessary in order to shape a future sustainable and resilient food system.

8.3.2 Community-Led Projects

The impact of educational activities – perhaps the strongest motivation in the sample of case studies – was generally considerable. As noted, the report produced by Real Food provided the number of people reached through their initiatives, including their aquaponic unit (e.g., in 2017, 832 students attended courses in the geodome). The consolidation of a model of educational provider, using food projects and soil-less technologies as educational tools, is evidence of its success. A glance at the websites of Sow the City and Die Urbanisten, and the several food growing projects showcased, shows that these non-profit associations not only reached a considerable number of schools and community groups but also that these

Fig. 8.1 Hydroponic unit included in the rooftop community garden in Manchester

clients were prepared to invest, believing in the value of such projects. This educational model is now quite widespread and other organisations can be found across Europe providing educational services. Bio-T-Full[3] for example, is an organisation promoting urban agriculture in the Nantes region, which has built hydroponic units in primary and secondary schools. FarmUrban[4] is a UK based organisation promoting vertical farming. They installed an aquaponic unit in a Technology College in Liverpool. The associations of the case studies agree that technology is a great pull for students. Interestingly, these educational projects can generate an impact that goes beyond the users targeted. A case in point is the hydroponic unit built by Sow the City for Manchester – European City of Science – which was later purchased by Printworks and relocated on the rooftop of the office building where this company operates, and subsequently transformed into a community garden. The manager of the garden regularly hosts groups of people and students, who show great interest in the hydroponic unit (Fig. 8.1). The recurrent question in this book about the impact that a non-natural method of food production used as an educational tool has on the perception and understanding of nature generally, is perhaps answered in a video on social media, presenting a rooftop hydroponic farm in Singapore, Orchard Street.[5] In the video, the farmer answers this question by stating that technology, hydroponic technology, does not solve anything. It is people and the way they utilise this

[3] http://bio-t-full.org/

[4] https://farmurban.co.uk/projects/

[5] https://m.facebook.com/story.php?story_fbid=1649538435145651&id=836689413097228

technology as a tool – and for a purpose – that will turn technology into something meaningful.

Another relevant outcome that can be identified in these projects is the active and fertile collaboration between academia, civil society and local governments. This collaboration shares similarities with the tetra helix models, proposed as an advancement of the triple helix model, conceptualising an active collaboration between academia, business and government in order to foster economic and social development (Etzkowitz & Leydesdorff, 1995). The tetra helix model adds civil society as a fourth element, necessary to ensure correspondence between economic and social development needs, and as an approach to democratising science and technology (Russell, 2015). Huerto Lazo is connected with the University of Sevilla and has hosted some training activities funded by the Spanish Minister of the Environment, organised by the association Aula del Mar[6] in Malaga, aimed at farmers willing to start an aquaponic commercial activity. A session of the training course (joined by the author of this book during the trip to visit Huerto Lazo) was attended by a varied audience, including farmers with prior experience in aquaculture and young unemployed. The audience also included a participant who had just completed a PhD in biology and, rather than starting an academic career, was keen to apply his scientific knowledge to implement an aquaponic farm. Generally, attendants showed previous knowledge of aquaponics. Their questions to the instructors were competent, focusing on specific technological and biological aspects. This experience further confirms that the level of technological knowledge of non-experts can be considerable and that the socio-economic profile of the new generation of urban farmers is highly varied and includes people with high academic qualifications.

Through these institutional connections, the training provided by Huerto Lazo is rather strategic, in that it promotes them as experts on a par with academic institutions (teaching in these training courses was delivered by lecturers from the University of Seville), while giving them access to institutional funding for economic development. This partnership (policymakers, research institutes and civic society) has great potential to create synergies and enable dissemination and application of scientific research in practice, benefitting community groups. With a different approach, Real Food and Sow the City collaborate with the University of Salford, which helped monitor the impact of the activities of Real Food and developed a tool with Sow the City to identify the scale of investment needed, and the return for enterprises starting soil-less farms (see Chap. 6). As mentioned in Chap. 1, LEAP (a small enterprise developing urban anaerobic digesters) is collaborating with academics and a community garden to develop a circular economy small scale food production system. Another interesting case in point of a tetra helix application is the Crop Cycle project, recently funded by the Welsh government.[7] The project is organised by Social Farms & Gardens (the UK national association of community

[6] https://www.auladelmar.info/aula/acuaponia-2

[7] https://businesswales.gov.wales/foodanddrink/news-and-events/news/crop-cycle-welsh-government-fund-supply-installation-controlled-environment-agriculture-systems

gardens and city farms) in collaboration with industry partners. In this project, four local communities will run small vertical farms unit, in retrofitted, fully equipped shipping containers, testing business models customised to the local context (Social Farms & Gardens, 2021). These collaborations are vital to ensure that new food technologies are developed in line with people's expectations and grafted effectively into urban communities.

The third relevant impact has clear political connotations. Milagro de los Peces focuses on education and self-supply, the latter specifically with its first prototype, located in the backyard of a household in Poligono Sur. This is an experiment in line with the concept of frugal innovation, which customises the FAO aquaponic prototype, designed for developing countries. The share of low-income households is higher in the southern and eastern areas of Europe and much smaller in the north and north-west areas (Eurostat, 2021). In a European, developed countries context, with an established social welfare system, self-supply for subsistence seems a remote option. But the rise in food banks not only in the UK and US but also other developed economy countries (Bacon & Baker, 2017) may easily lead to more households deciding to run an aquaponic backyard unit. Also, some of the 'YouTubers' in Chap. 7 showed aquaponics farmers adopting this technology to survive in extreme environmental conditions of countries such as Australia. Backyard soil-less systems may become a desirable option not only because of economic crises but also because of climate change.

It is worth noting that self-supply is envisaged as a market opportunity targeting middle-income households too. Citybotanicals, a German organisation providing consultancy on the integration of food growing within urban green infrastructure, has engineered and sells an aquaponic greenhouse at a price of about €23,500.[8] Self-supply is in line with much of the urban gardening activities at present happening in Europe either in allotments or in home gardens, the latter representing the largest urban land use type (Ghosh, 2014) and being cultivated with edible crops in many cases (Galluzzi et al., 2010). Self-supply in home gardens is therefore widespread and, by extension, changing environmental and social conditions may result in a shift from crops grown in-soil to soil-less units in private gardens, in the not-too-distant future. These experiments with self-supply are therefore an important step forward to trial models and technologies and have them ready for implementation when this becomes necessary.

8.4 Productivity

Before we attempt to quantify the level of productivity across the case studies some clarifications are necessary. The first one relates to the extent to which productivity matters to, and is understood by, the urban farmers interviewed. For some of them,

[8] https://citybotanicals.com/shop?category=Komplettsystem

as noted, the attainment of social objectives is a priority. Therefore, a definition of productivity based on quantities of food harvested does not capture the scope of these projects, which goes well beyond the food produced. It excludes benefits connected with social betterment, which are difficult to quantify and therefore appreciate. Food is a complex issue, with profound repercussions on the environment, people's health, and local and global political stability. Interacting with all these facets at once is one of the most valid merits of urban agriculture, including these soil-less projects. Appraising them by the quantity of food produced diminishes their real impact. In fact, more research is needed to generate an evidence base that can demonstrate the level of contribution to society of urban agriculture projects generally. A case in point is a pilot study based on a London community garden, which uses a simplified Cost Benefit Analysis to ascertain that a return on investment of £3 to £1 is generated when activities related to both food and improvement in mental wellbeing of volunteers are considered, with the latter showing the highest share of return (Schoen et al., 2020). Another study measuring harvest and social benefits in a London community garden shows that produce harvested is relatively low and social benefits perceived by volunteers high, with these two factors possibly correlated (Caputo et al., 2021).

The second one is that measuring productivity in terms of food produced per unit (e.g., kg/m^2 or € invested per kg harvested) requires benchmarks, which for soil-less technologies are quite difficult to establish because of the paucity of literature documenting their productivity and, more importantly, productivity referred to small scale projects as those included in the book. These projects cannot be compared with large scale commercial projects, which are technologically advanced and employ experts for their management. In the literature reviewed for this book, productivity is usually defined for a few crops only. In the case studies, a range of crops is produced, and disaggregated data are not available. Rakocy's farm at the University of Virgin Islands (see Chap. 4) was designed in order to establish the size of a commercially viable unit, therefore in line within a commercial aquaponics logic. The Virgin Islands farm did not utilise sophisticated greenhouses and artificial lighting, and its size was contained; it functioned as a testbed to identify optimal factors for maximised production and resource efficiency. Today, hydroponic and aquaponic companies may reach a higher level of productivity compared to the Rakocy's experiment, supported by advanced technology and equipment that, however, is not available for small projects. A review by Barbosa et al. (2015 – see Chap. 4, Table 4.2) of relevant literature, established that, with a density of 24 lettuce heads per square metre, 12 harvests per year and an average weight per head of lettuce of 0.145 kg, the annual yield of lettuce in a hydroponic greenhouse in a climatic zone similar to that of Arizona, US, is 41 kg/m^2/year. But types of crops and crop density can influence harvest. In Table 4.3, Chap. 4, the yield reported by Rakocy is quantified using the averaged lettuce weight used by Barbosa et al. and that reported by Rakocy, which is much higher. When using the latter, the yield is 50.88 kg/m^2/year; with the former, the yield is 27.84 kg/m^2/year. Our case studies often grow leafy greens and microgreens, with low-volume, high value yields. Crop densities are between the two studies mentioned. In Fig. 8.2 both studies (i.e., Rakocy and

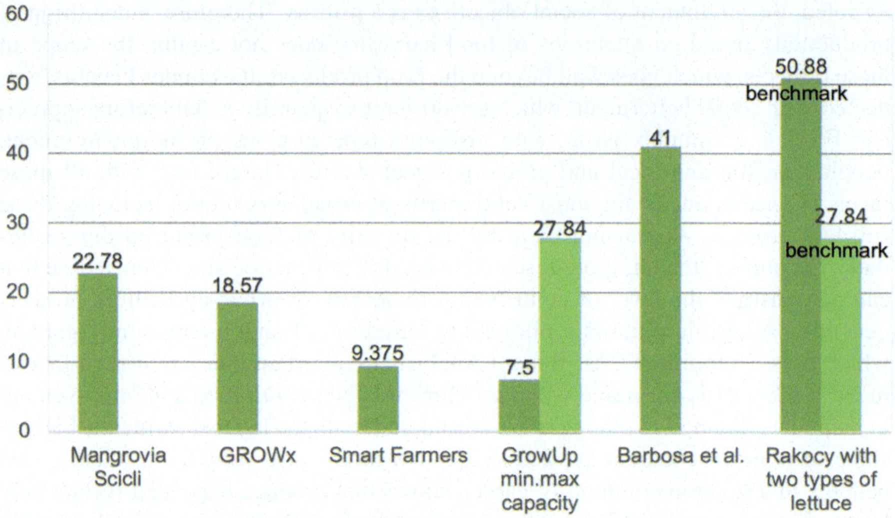

Fig. 8.2 Productivity of leafy greens (kg/m^2) in some of the case studies, compared to productivity benchmarks in the literature

Barbosa et al.) are reported as benchmarks. In Table 8.2 the productivity of all our case studies, hydroponic and aquaponic projects, is summarised. This is because only one small scale hydroponic farm is included in the case studies, with the other hydroponic projects utilised for educational purposes or community building (Hemmaodlat and Sow the City).

Another clarification relates to the reliability of the data collected. These were provided by the soil-less farmers, and they may be broad estimates or, at best, approximations of the actual yield. An exception is Milagro de los Peces, which accurately recorded inputs and outputs, documenting them in a study that compares two similar aquaponic units, demonstrating similar levels of yield (Lobillo-Eguíbar et al., 2020). Generally, projects such as Real Food are not equipped and organised to collect data and as noted, productivity is not their main objective. Small enterprises such as BioAqua and Mangrovia Scicli would normally keep detailed records of their production, but they may not have sufficient time for comprehensive data collection or may be reticent in providing an in-depth picture of their technological and commercial performance. Medium size enterprises, such as GrowUp, may not disclose information deemed as sensitive. As a result, Table 8.2 presents many gaps and is incomplete. More importantly, the size of the sample of case studies is not sufficient and congruent to be representative of small-scale, urban soil-less real productivity. Rather, it is a demonstration of the great variability in approaches to production and the way production itself may be substantially affected by social and organisational factors. Real Food, Hemmaodlat and Huerto Lazo, for example, do not focus on production but on education, with the former clearly stating that

Table 8.2 Size and productivity of the case studies

Case study	Production units and area	Crop harvest	Crop productivity	Fish harvest	Fish productivity kg/m^2/year (kg/l/year)	Crop variety	Fish variety
Milagro De Los Peces –	1800 L (fish tanks) + 4.56 m^2 (hydroponics)	177.66 kg/year	38.96 kg/m^2	33.5 kg/year	7.44 kg/m^2 (0.0186 kg/L)	22 fruit and vegetables	Red hybrid ttilapia
Huerto Lazo –	135,000 L (fish tanks) + n/a (hydroponics)	16,000 plants		1200 kg/year	0.00889 kg/l/year	Bhrami, jiaogulan	Tilapia, catfish, tench
Real food –	2200 L (fish tanks) + 5.5 m^2 (6 m tube +4 m^2) (hydroponics)	100 kg/year	18.18 kg/m^2	50 kg/year (estimated)	1.11 kg/m^3 (0.0227 kg/l)	Leafy crops	Tilapia, catfish, crayfish
Hugh –	1000 L (fish tanks) + 5 m^2 (hydroponics)	Not recorded		No consumption		Mint, chive, lettuce, tomatoes, water cress, swiss chard, cucumbers, herbs	Grass carp, tench
Mangrovia Scicli –	60,000 L (fish tanks) + 700 m^2 (hydroponics)	100,000–110,000 plants/year	22.78 kg/m^2	1500 kg/year	2.14 kg/m^2 (0.025 kg/L)	Chard, lettuce, endive, chicory and other leafy greens	Largemouth bass
Bioaquafarm–	40,000 L (fish tanks) + 500 m^2 (hydroponics)	Not recorded		1200 kg/year	2.4 kg/m^2 (0.024 kg/L)	39 different crops[a]	Trout
Smartfarmers–	6000 L (fish tanks) + 48 m^2 (vertical hydroponics)	450 kg /year (leafy greens)	9.375 kg/m^2	600 kg/year	12.5 kg/m^2 (0.1 kg/L)		
Growx –	100 m^2 (vertical hydroponics)	52,000 punnets/year	26 kg/m^2			Microgreens	

(continued)

Table 8.2 (continued)

Case study	Production units and area	Crop harvest	Crop productivity	Fish harvest	Fish productivity kg/m^2/year (kg/l/year)	Crop variety	Fish variety
Growup –	36,000 L (fish tanks) + 750 m^2 (vertical hydroponics)	Designed for 20,000 kg/year. Produced 37,500 bags salads/year	Optimal productivity 26.66 kg/year Produced 7.5 kg/year	Maximum tank capacity 200 fish (12 tanks available)		Leafy green salads, herbs, microgreens	Tilapia
Mushroom farms							
Rotterzwam	236 m^2	18,000 kg/year					
Treuffelwerk	100 m^2	3000 kg/year					

[a]rocket, shiso, amaranth, parsley, coriander, basil (five varieties), spring onion, chard, samphire, capsicum peppers, chillies, tomatoes (six varieties), cucumber, aubergine, cabbages, kale, pak choi, mibuna, celery, courgette, grapes, fennel, salad, dwarf bean, green peas, mange tout, corn, parsnips carrots, celeriac

revenues coming from their social support activities to young people are a funding source for their aquaponic project.

Comparability is also problematic. In each case study different baskets of crops are grown, but data disaggregated per crop are not available. However, many farms grow leafy greens and microgreens only, as these are fast growing, high value crops. The cycle of lettuce can be as fast as 4 weeks from seedling transplanting to full growth. Potentially, it can be harvested 12 times per year, with an estimated total yield between 27.84 and 50.88 kg/m^2 per year, depending on the type of lettuce grown (see Table 4.3, Chap. 4). These two yields are used as benchmarks in Fig. 8.2. Optimal conditions are fundamental for such a high yield, with correct high plant density, nutrient solution and ambient climate all contributing. The yield mentioned above refers to the University of Virgin Islands farm (Rakocy et al., 2011), where these factors performed optimally. In colder climates and, for example, with a market demand for crops that are less productive in terms of weight, the yield can vary substantially. A case in point is the study on the three aquaponic farms reported in Chap. 4, Table 4.4 (Tokunaga et al., 2013), with two aquaponic farms growing lettuce only, producing respectively 24.8 and 77.66 kg/m^2. In this study, productivity was not calculated on the growing surface area but on the total surface area of the hydroponic farm. Although not accurate because of the metrics used, the difference in yield is notable and demonstrates the difficulty of establishing benchmarks.

With data aggregated such as those provided by Milagro de los Peces, productivity cannot be easily determined: different combinations of vegetables and fruits can result in higher and lower productivity, depending on the choice of crops. For example, aggregating different crops included in Chap. 4, Table 4.1, yield varies from 29.7 to 49.3 kg/m^2. For the case studies that have quantified their productivity in plants harvested (e.g., Mangrovia Scicli), the weight of a head of lettuce as given by Barbosa et al. (2015) was considered (0.145 kg), which is rather low and therefore comparable to their crops (leafy greens). The most productive is Mangrovia Scicli, with a yield of 23.57 kg/m^2. GROWx comes next, with 18.57 kg/m^2 where a weight of 50gr for each punnet of microgreen sold is considered. Smart Farmers and GrowUp are below 10 kg/m^2. GrowUp's productivity was calculated assuming 150 g for a bag of salad. At the time of the visit to GrowUp facilities, the vertical farm and the fish tanks were not used at full capacity, which, when reached, would result in higher yields. As stated on their website, the farm was designed to produce 20MT/y. When this is considered, GrowUp is the most productive farm, reaching 27.84 kg/m^2 of high-value produce. The fact that GROWx produces less than GrowUp's expected yield is perhaps connected to the low-weight / high-value crops that the former grows. They both reach or almost reach the lower bracket of Rakocy's benchmark, which is the most realistic term of comparison, considering the type of low-weight crops both enterprises grow. However, GROWx and GrowUp are vertical farms; their space efficiency should result in higher productivity measured in kg/m^2. Mangrovia Scicli's productivity is slightly below GrowUp's expected final yield, thus showing effective use of the deep water culture technology. The lower yield of Smart Farmers can perhaps be partially explained by the combination of low reliability of the estimated crop harvest, or the manager's level of farming expertise.

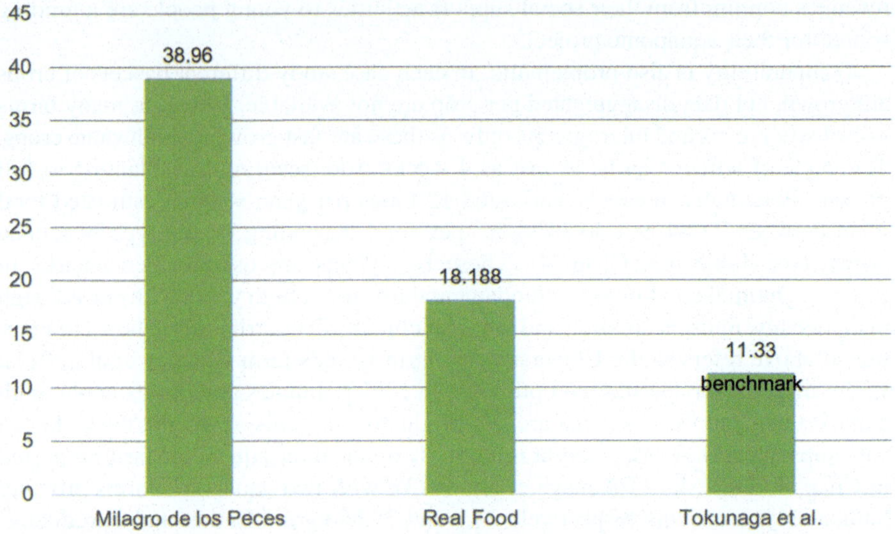

Fig. 8.3 Productivity of mixed crops (kg/m^2) in some of the case studies, compared to productivity benchmarks in the literature

With multi-crop farms such as Milagro de los Peces and Real Food, the comparison with the third industrial farm in the Takunaga study (Chap. 4, Table 4.4) shows different results (see Fig. 8.3). The industrial farm grows 11.33 kg/m^2, almost as much as Real Food (18.18 kg/m^2) and much less than Milagro de los Peces (38.96 kg/m^2). Here too, data are not completely reliable. Takunaga's study on the industrial farm does not provide the cultivated hydroponic surface area but only the area of the farm, therefore the real yield should be higher. Real Food provided an estimate, which is not based on data collected regularly but on an evaluation of the chief gardener, which is however probably closer to reality. But the yield of Milagro de los Peces is accurately documented. The high productivity of this case study suggests that the backyard aquaponic model can be very successful. Small household units can produce large quantities compared to the space occupied by their small unit, possibly as a consequence of their motivation for growing food (food security) and the time and dedication allocated to it. Milagro de los Peces shows that this backyard model, when replicated in a community environment, can be equally effective.

Some of the case studies may be more productive in terms of value generated, rather than volume. Microgreens are high value crops, generating higher returns compared to many other crops. In farms where the range of crops cultivated is wider, such as BioAqua Farm which grows 39 different crops, these are all high value crops too (e.g., herbs and samphire). Generally, the choice of high value and fast-growing crops is particularly appropriate for small scale soil-less urban projects, as quality rather than quantity of produce ensures sufficient income. A drawback can be that such a selection of crops can be nutritionally high and yet low in calories and proteins, not including staple crops that are fundamental to any diet.

Pulses, for example, offer high protein content but lower returns, require more space and care, and have a slower growing cycle than leafy greens. This raises questions about the suitability of crops typically grown hydroponically in small scale urban soil-less commercial projects, from the perspective of food security. While projects such as Milagro de los Peces offer a model of backyard aquaponics that is suitable to strengthen food security, the commercial logic of small for-profit projects works in the opposite direction, even when these are socially oriented. In a study developed on a sample of 68 aquaponic farms (Villarroel et al., 2016), herbs, lettuce and tomatoes were by far the three most grown crops (respectively by 60, 50, and 35% of all respondents). The share of crops and fish cultivated, which in commercial projects is generally higher for crops and smaller for fish, changes in Milagro de los Peces to 52.3% and 47.7% respectively, when measured in terms of economic value. In this project, food grown and harvested meets the needs of a complete diet, which includes proteins, fibres and other nutrients (Lobillo-Eguíbar et al., 2020).

Mangrovia Scicli, the only case study in which directors have a higher education qualification and experience relevant to their aquaponic enterprise, is the most productive within the sample of case studies. Although the directors, admittedly, are still learning, they have prior experience that the other farmers cannot claim. The type of knowledge and the way this is transferred are important factors that, generally, distinguish urban farmers from soil-less urban farmers. The former can rely on horticultural science, transferred through training courses or by other farmers while practicing; there is a long tradition and a large community of expert urban farmers that can transfer knowledge intergenerationally. The latter is operating on new grounds, with knowledge and experience that still needs to be built and consolidated. Soil-less urban farmers cannot learn from previous generations, but rather only by doing and, as we have seen in the Internet case study in Chap. 7, by utilising the web and other digital tools such as YouTube videos. It is understandable that, for non-professional soil-less farmers, high productivity can be difficult to attain.

In Fig. 8.4 (see also Table 8.2), the fish productivity is quantified in kg/m^3. The benchmark used for this chart comes from Racocy's experiment at the University of the Virgin Islands aquaponic farm, which attempted to maximise production and demonstrate the potential of the aquaponic technology for industrial use. The resulting productivity levels were high and in line with this aim. The fish productivity measured in the case studies reflects different strategies and aims, although not always in a clear way. An important factor influencing their productivity is the business model which can prioritise crop rather than fish production, or vice versa. Many soil-less farmers declared that crop sales generate higher revenues, and their farm is designed to increase crop rather than fish production. Rakocy estimated that his farm could harvest annually between 4160 kg for Nile tilapia and 4780 kg for Red tilapia (Rakocy et al., 2011). Using Nile Tilapia, Rakocy's farm could produce 133.33 kg/m^3, which is the benchmark used in the chart. Milagro de los Peces (98.6 kg/m^3) and Smart Farmers (100 kg/m^3) are the case studies with the higher productivity, although the former is recorded on the real harvest and the latter is a projection. Milagro de los Peces is an educational project designed as a self-supply aquaponic unit, hence attempting to demonstrate the potential to reach a good level

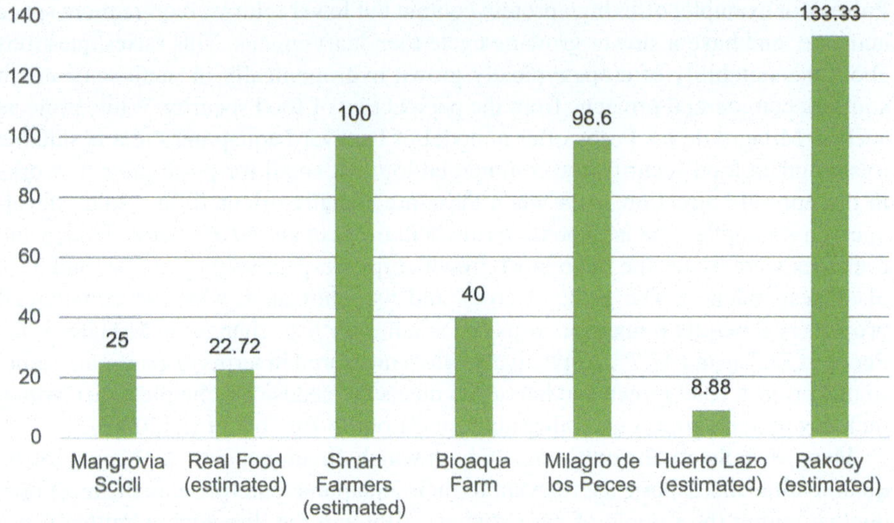

Fig. 8.4 Productivity of fish (kg/m³ water) in some of the case studies, compared to productivity benchmarks in the literature

of self-sufficiency in food production. Smart Farmers aims to demonstrate the potential for food production and is experimenting with different types of fish to increase harvest and revenues. BioAqua Farm and Mangrovia Scicli show rather low and similar level of productivity (respectively 40 kg/m^3 and 25 kg/m^3). For both farms, the sale of crops represents the major source of income and it can be supposed that the fish population is low but sufficient to provide the nutrient necessary to the hydroponic unit. However, they both mentioned that the fish sale is part of their business model. The gap between their productivity levels and those of Milagro De Los Peces and Smart Farmers is difficult to explain. The productivity of Huerto Lazo is very low, although this is not surprising, considering the organisation of the project. The productivity of Real Food is close to that of Mangrovia Scicli, although their fish tanks are very small and the harvest is a rough estimate.

8.5 Environmental Performance

As for productivity, for similar reasons, establishing the efficiency of resource use for projects at this scale is not easy. Data may not be completely reliable, and the small sample of case studies does not allow for any generalisation of findings. Many studies on aquaponics and hydroponics point at energy use as the factor that can significantly increase the environmental impact of these technologies when compared to in-soil production (Chen et al., 2020; Romeo et al., 2018; Okemwa, 2015). Infrastructure needed for greenhouses and the specialised equipment is also a source of environmental impact (Romeo et al., 2018; Benis et al., 2015). Data collected here and the analysis developed are not sufficiently detailed, and only point to great differences in performance between urban small scale and industrial soil-less farms. The question is whether with considerably inferior means and knowledge, small scale, community-led projects can compare with professional ones. Here too, appropriate benchmarks are difficult to identify. In the University of Virgin Islands project, on average, added water was 1.5% of the tanks' volume daily and 2.21 kWh of electricity was used (Rakocy, 2012). Virgin Islands' climate is tropical, with an annual mean temperature of 28/29 °C. This may result in high levels of evaporation of the water in the tanks and water intake from plants. At the same time, fish tanks in the cold months did not need heating and the aquaponic farm did not use artificial lighting. For fish feed, Rakocy identified 57 g as the optimum ratio of feed weight per m^2 hydroponics cultivated area, with lettuce as the only crop grown. How environmentally efficient these benchmarks are and to what extent they can be advanced it was not possible to establish. Debate on the environmental efficiency of aquaponics and hydroponics, compared to conventional agriculture, reports different findings. Barbosa et al. (2015) compared conventional and soil-less methods to determine differences in water and energy use for lettuce production only, in Arizona. The study concluded that water consumption is comparable between the two methods in terms of surface area, but hydroponics is about 13 times more water efficient when yield is considered, with conventional lettuce using 250 ± 25 L/kg/year and hydroponic lettuce using 20 ± 3.8 L/kg/year. The study is based on an average weight of 144.6 g per lettuce head.

Resource use of the case studies is reported in Table 8.3. Milagro de los Peces is in line with Rakocy's results of water added on a daily basis (1.5%) whereas BioAqua Farm is more virtuous (1%) and Smart Farmers more profligate (1.97%). Mangrovia Scicli adds between 1.5 and 3% of the volume of the fish tanks. Real Food does not seem sufficiently reliable in recording water use. It can be concluded that at a small scale, Rakocy's findings hold true, and that variations in the case study sample are attributable to different crops cultivated, local climate, equipment and individual approaches to the management of the farm. Establishing the efficiency of water use in relation to production of crops is more difficult. Barbosa's study offers a benchmark for lettuce (20 L/m^2 – Chap. 4, Table 4.2). Assuming that crops and productivity of Smart Farmers can be assessed against this benchmark, and assuming that the water added daily is mostly absorbed by crops (70–80% for crops and 20–30% evaporated or lost), 1 kg of lettuce grown would require between 49.7 and 56 L; almost three times the benchmark and yet one fifth of the average consumption for conventional agriculture. GROWx, the only hydroponic farm included in the case study sample, did not provide a precise quantity for water use. If a fair interpretation of the description of water consumption provided by the company (a few m^3) is within the range of 5–10 m^3, then the crop production requires between 23 and 46 L/kg over the yearly production (Fig. 8.5).

This lower volume of water use (when compared to the other case studies) can be justified by the state-of-the-art equipment of GROWx, utilising a computerised management system and an indoor environmental control system.

Table 8.3 Resource efficiency of the case studies

	Size	Water added daily	Water for irrigation	Energy consumption – daily	Onsite energy generation	Fish feed
Milagro De Los Peces	1800 L (fish tank) + 4.56 m^2 (hydroponics)	27.15 L 1.5%		0.95kWh (334.5 kWh/year)	Thermal solar to heat water in fish tank	45.5 g/m^2/day
Huerto Lazo	135,000 L (fish tanks) + unknown (hydroponics)			50kWh (18,000 kWh/y)		15 kg/day
Real food	95 m^2 2200 L (fish tank)/6 m tube +4 m^2 (hydroponics)	5 L/day 0.22%		N/A	1 PV panel installed	Not known
Hugh	1000 L (fish tank) + 4–5 m^2 (hydroponics)			0.3 kW (pump)		Not known
Mangrovia Scicli	60,000 L (fish tanks) + 700 m^2 (hydroponics)	1000/2000 L/day 1.6–3.3%		4 kWh (1408 kWh/year)		25–50 g/m^2/day
Bioaqua farm	40,000 L (fish tanks) + 500 m^2 (hydroponics)	142 L/day avg. 0.3%		72kWh (26,280 kWh/year)	Ecotricity	40 g/m^2/day
Smart farmers	6000 L (fish tanks) + 48 m^2 (hydroponics)	118 L/d purified wastewater 1.97%		88.5 kWh (32,300 kWh/year)	2 kW peak	42.6 g/m^2/day
Growx	100 m^2 (vertical hydroponics)		a few m3 month	833 kWh (300,000kWh/year)		
Rotterzwam	Mushroom farm			123 kWh (45,000 kWh/year)	50,000 kw/year	

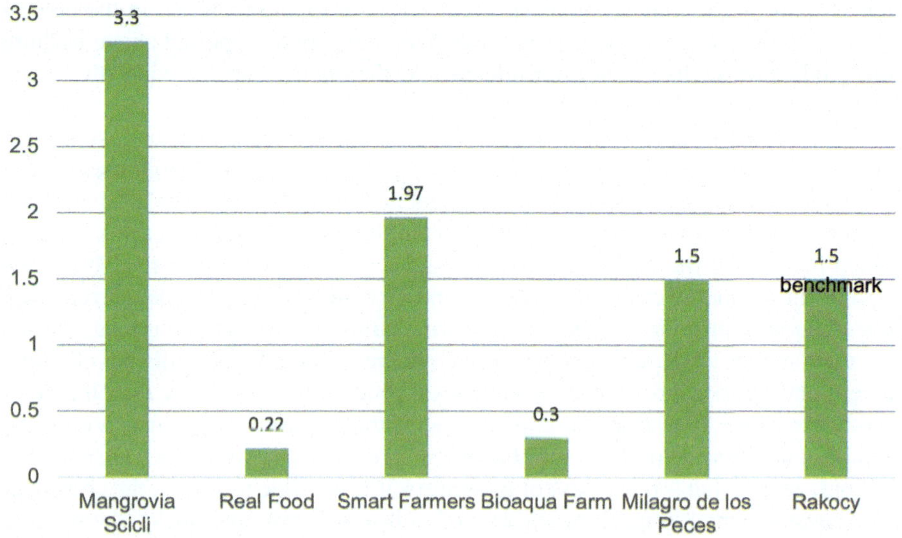

Fig. 8.5 Percentage of existing volume of water added to the tank each day in the case study soilless systems

According to the data provided by GROWx, with 50 g as an estimated weight of a microgreen punnet, the energy deployed per unit of food grown is very high (115 kWh/kg) when compared to the estimate by Barbosa et al. (2015) of 25 kW/kg for hydroponics (see Table 4.2, Chap. 4). However, GROWx functions with artificial light only and it is digitally controlled, whereas with the report of Barbosa et al., energy consumption was measured in a greenhouse in Arizona, benefitting from solar energy, with monthly average temperatures peaking 34.7 °C in summer and as high as 14.1 °C in winter. Smart Farmers records 2.7 times more energy consumption (68.9kWh/kg/year) compared to Barbosa et al. Like GROWx, they grow indoors (although not with a vertical farm system) and with artificial light. Also, the focus of the company is the design of a marketable aquaponic model; it is likely that with intensive production of crops, this company would consume more energy too. Smart Farmers includes onsite renewable energy production, which can yield about 1824 kWh/year.[9] This energy is generated by 1 panel, with a peak of 2 kW. When this is considered, the energy necessary to produce 1 kg of leafy vegetables is reduced to 65kWh/y. It would take a relatively affordable investment (5 PV panels) to further reduce the electricity load of Smart Farmers to 49.5kWh/kg. With less sophisticated equipment, no artificial lighting and the hydroponic unit located in greenhouses built with polyethylene membrane and steel frames, Mangrovia Scicli consumes much less energy; a fraction of the average energy for 1 kg of lettuce reported by Barbosa et al. at between 0.075 and 0.085 kWh. In Scicli, the average

[9] https://www.pvfitcalculator.energysavingtrust.org.uk/

temperature in the hottest month is about 30 °C dropping to 15 °C in the coldest month. The real energy intensity is lower than the quantity reported above because the production of fish is not considered and the energy use for aquaculture and hydroponic has not been disaggregated.

BioAqua Farm and Milagro de los Peces are not included in this assessment because they grow a wide variety of crops, but, for the latter, the energy used to heat the water tank in the coldest months is generated through solar panels. The energy used for other operations is 334.4 kWh/y, resulting in 1.94 kWh for each kg grown, much lower than the energy necessary to grow lettuce conventionally. This is the best result within the case study sample, and it could be brought to below zero with the addition of just one PV panel to the farm equipment. BioAqua Farm has chosen for their supply a UK renewable energy company, paying higher bills but claiming a zero-energy food production. RotterZwam produces more energy than they currently consume. It is also the company with the highest energy load within the sample, and the scale of the renewable energy production to which they are committed only shows that – regardless of the scale – it is possible to reduce the conspicuous environmental impact of energy of soil-less farms and make them more competitive with conventional agriculture on this front. The initial investment necessary for onsite renewable energy equipment is a barrier for many of the small community-led groups and small enterprises, but this may soon change with cheaper and more efficient renewable energy components and, hopefully, higher commitment from central governments to provide much needed incentives to increase the uptake of renewable energy generation units.

These sections of the book are meant to provide an overview of productivity and environmental performance of the case studies. Comparisons between case studies and benchmarks are provided but they should be read with caution. It is more useful to observe how, in the face of considerable constraints on investment, equipment and know-how, the case studies attain notable achievements. One of the conclusions of this analysis is that motivations play an important role in this, with self-supply perhaps being one of the strongest and most effective motivations, as Milagro de los Peces shows. Comparisons are illustrative also because of the scale and scope of the farms that were used as benchmarks. Most of the case studies are small, built with basic components and designed to be low cost and low consumption. For them, good productivity and environmental performance should be harder to attain. Instead, some farms suggest that the opposite is true, which leads to the conclusion that a critical mass of small soil-less urban farms can generate many more direct and indirect benefits than large scale soil-less farms.

Chapter 9
Conclusions and Future Steps

Abstract This chapter traces possible futures for small scale soil-less urban agriculture, building on an initial broad view of the factors contributing to the emergence of this new phenomenon, the global context of food production and supply chains, and findings from small scale soil-less urban agriculture case studies across Europe. The propensity of urban farmers to experiment with new social, physical and practice-based arrangements, the relationship between people, science and technology, and food, and the stories of the farmers and projects recorded in the case studies are pieces of a puzzle that is not complete and will probably mutate in the near future, when soil-less urban agriculture will further grow and consolidate. Following the initial sections elaborating on technology and food production, the last sections identify and discuss the key contributions that small scale projects can give to the soil-less sector, which are instrumental to avoid the replication within the high-tech food sector of unsustainable practices characterising industrial in-soil agriculture. Natural versus unnatural, small, versus big, rewilding versus farming are some of the dichotomies that need to be resolved for soil-less urban agriculture – particularly at a small scale – to expand.

The purpose of this last chapter is to conclude the book by tracing possible futures for small scale soil-less urban agriculture, building on the investigation presented here. The book provides a broad view of the factors contributing to the emergence of this new phenomenon: the propensity of urban farmers to experiment with new social, physical and practice-based arrangements, which generate innovation; the relationship between people, science and technology, and food; the historical trajectories of three soil-less technologies; and the case studies capturing the diverse facets of a trend that is small but expanding, are all pieces of a puzzle that is not complete and will probably mutate in the near future, when soil-less urban agriculture will further grow and consolidate. In fact, the book is an exploration of trends that are quickly unfolding. In order to draw up a short-term future scenario for soil-less small-scale urban agriculture, this chapter will review some of the topics presented in Chap. 1 that provide the context and the rationale to pursue alternatives to current food systems, proposing urban agriculture as a valuable component of such alternatives. Technology

S. Caputo, *Small Scale Soil-less Urban Agriculture in Europe*, Urban Agriculture, https://doi.org/10.1007/978-3-030-99962-9_9

and food technologies will be discussed once again; and finally, some possible pathways to expand soil-less agriculture in cities will be explored.

9.1 Context

The environmental crisis represents a turning point for society. Its consequences are so deep and dramatic that only bold, radical transformations of the way we live can lead to containing our footprint within the ecological boundaries of the planet. The transformation of food systems is arguably one of the key areas requiring urgent changes in food production, policy, land use and food culture. Food systems cross diverse sectors and scales, from cities to regions and the entire globe; any radical change will require complex solutions with potentially problematic processes of implementation. In an environmentally and socially unstable world, many studies point to increased self-sufficiency (Kriewald et al., 2019; Lang, 2020a, 2020b; Pradhan et al., 2014) and food sovereignty (García-Sempere et al., 2018) as necessary conditions to strengthen local resilience. But these conditions clash with a current state in which food production dynamics are geographically uneven in terms of quantity, quality and value. For example, globally, an estimate shows that between 1961–2010 cropland expanded by 0.24% per year on average but only 0.04% per year from 1995–2010 (Ausubel et al., 2013). The expansion is uneven, with the 'agricultural frontier' over the last decades expanding towards Latin America, Africa and Southeast Asia, and a contraction of cropland happening elsewhere (Ramankutty et al., 2002, 2018). Europe is experiencing a contraction of farming, especially on marginal land, with some projections suggesting that, by 2030, more than 30 m hectares of farmland will have been abandoned (Tree, 2018:154). In the USA area of cropland fell by 3% from 23.7% in 1970 to 20.7% of total land use in 2012 (Spangler et al., 2020). However, in Brazil, cropland increased by 24% from 65.37 to 81.18 million hectares during the last 10 years (Statista, 2020b). Changes in land use are determined by global market dynamics, with local exploitable resources, national policies and cost of labour playing a decisive role. Transforming this seems an immense endeavour, especially considering that the agricultural industry is strongly linked with other powerful players such as chemical and biotech sectors, which can influence the market and the policy. Moreover, although acknowledging the externalities of the agricultural industry, policy is resistant to change. EU agricultural policies (CAP) reward land ownership regardless of its use and there is a reluctance to change this system.

9.2 Technology

A driver of change is technology. Technology applied to agriculture has greatly improved productivity. Historic trends show that from the 1960s yield improvements account for 77% of increased production (Ramankutty et al., 2018). These higher yields are a consequence of higher input rates and technical change, calculated as the ratio between total outputs of crops and livestock to total aggregation of all the land, labour, capital and materials used in production when these save costs (Fuglie & Wang, 2012). Recently, GPS and remote sensing technologies have been used to develop *precision agriculture*, a term describing a technology-based approach to minimise resource use by identifying cropland areas requiring inputs, using them in these areas only (Bramley, 2009). Precision technology is still expensive, but the key point is that research and development is a critical factor for a transition to a sustainable agriculture and food system generally. As noted, technology is not neutral and its direction is determined by varied interests, including those dictated by mere financial returns. Technology is ever-more present within the public food debate, both in developing alternatives to the current unsustainable food system and augmenting its unsustainability. Food technologies, for example, have been applied to animal nutrition, sometimes with unintended and dangerous outcomes such as risks of infections (e.g., salmonella and E. Coli), ingestion of chemicals and diseases (e.g., BSE and Avian flu), which were widely debated in the media from 1990s onward (Blue, 2010).

Food science today is not only developing solutions for higher food safety and longer shelf life, but also for its enrichment by adding nutritional values as well as new textures and flavours. Processed food – from cornflakes to spreads – is produced utilising complex processes and technologies, although much of this food is nutrient-rich and unhealthy (McClements, 2019). But research and technology are also focusing on other 'future' foods which are protein-rich, can be used to reduce food's environmental impact and the production of which can be potentially scaled up to meet global demand (Parodi et al., 2018). A case in point is plant-based foods, which over the last years have attracted consumers' interest (Tziva et al., 2020). Protein ingredients can be extracted from crops, seeds, grass or seaweed (Aschemann-Witzel et al., 2020) and processed into food with specific textures, consistency, flavour and nutritional values comparable to meat, and could potentially substitute meat consumption with considerable reduction of GHG emissions linked to farming (Chai et al., 2019). A study suggests that customers perceive plant-based foods as 'cleaner' (free from chemicals) (Aschemann-Witzel et al., 2020), thus safer and healthier than industrial food.

Information provided and debated does not always help people to identify potentially beneficial food technologies, because it restricts such a debate to public health (i.e., food scares) without linking it to wider issues that are equally significant. Debate on soil-less produce is, as noted, a case in point. The use of chemicals and the perceived artificiality of the growing techniques are elements that can potentially hinder its uptake. These legitimate doubts do not hold in the face of the real impact of the entire conventional agricultural industry which is neither natural nor free from chemicals. The damage to biodiversity from current agricultural systems is so extreme that many believe that only rewilding policies incentivising conversion of marginal agricultural land into wild habitats (Land Sparing) can mitigate it (Monbiot, 2014; Tree, 2018). One of the issues highlighted by these advocates of rewilding is that the general public identifies the agricultural landscape as 'nature' rather than the result of centuries of human intervention. These landscapes have become a romantic vision of 'anti-modernity' (Tsing, 2015), where technology has no place, although they are the unintended result of at least two centuries of technological progress in agriculture, farming, and market exploitation. Returning nature to its original state may be impossible but restoring its real functioning is achievable only if we recognise the role food can play in this process and understand this process as situated within planetary, rather than local, limits. If food technology can help us to restore ecosystems functioning as well as to eat healthily, then it should be promoted. Indeed, the problem of framing the food discourse and communicating related sustainability issues to the general public is fundamental. Soil-less technology suffers from the same bias associated with technologies that have triggered diseases and pollution, although it is one that can generate ecological benefits. Yet its real value in terms of freshness and health, as Mangrovia Scicli shows, is difficult to communicate to consumers. The sustainability narrative that can convince them still needs to be developed (Junge et al., 2017).

9.3 Agricultural Industry

Technological progress has also determined the composition of the agricultural industry, providing large farms with equipment, products and techniques to organise cultivation over large areas, more efficiently and productively. As of 2000, a global estimate shows that there are more than 570 million farms occupying agricultural land worldwide. The majority of these farms is smaller than 2 hectares and family-run. Only 16% are larger than 2 ha, but they represent 88% of the world's farmland (Lowder et al., 2016). This estimate contradicts the claim that small farms produce the majority of crops globally (Wolfenson, 2013); it would be unlikely for this to happen on only 12% of cultivated land. An alternative estimate shows that small farms produce 28–31% of total crop production and 30–34% of food supply on 24% of all agricultural area (Ricciardi et al., 2018). These numbers demonstrate the great productivity of small farms, generating one third of food on one quarter of agricultural land, but they also demonstrate that the land ownership and the organisational and labour structure is increasingly in the hands of a few large farms and that the number of small farms is declining in developing countries (Lowder et al., 2016). Although fewer in number, large farms' agricultural practices in terms of pest control, fertilisers, tillage methods, number of crops cultivated and more will have the largest impact on the planet. If agroecology studies claiming that small farms are best placed to practice a more sustainable farming approach are confirmed (Altieri et al., 2015), a different pattern of ownership and market structure need to be developed, in which returns are more fairly distributed across the food chain, and policies facilitate small businesses to prosper. Agroecology also recognises the link between an approach to cultivation that respects the local ecology and the health of the soil, linking this respect to the social and political values of food in terms of right to food and food security (De Schutter, 2012).

Technology is an enabler of large scale farms in soil-less farming too. A report by Agrilyst (2016) suggests that 'indoor agriculture is one of the fastest growing industries in the USA'. Currently, the hydroponics market accounts for USD 7.9 billion. Estimates suggest that this market will more than double its size by 2025 (Agriculture Post, 2020). Estimates vary, with others being more conservative (e.g., an estimate of USD 9.8 billion by 2025 (Grard et al., 2020)), but they provide a plausible order of magnitude and, more importantly, indicate that a large share of this income will be generated by the technology and biotechnology companies providing equipment and software for large scale indoor farming. The capital cost to start indoor soil-less farms is high and the economic viability of these investments can be reached mainly through a sufficiently large scale of production, and only over a number of years. This limits the uptake of soil-less technology by small enterprises that cannot afford high capital investments and a long, initial phase before breaking even. It is a mechanism that favours large companies investing in this sector, with strong similarities to the farm ownership landscape outlined above, in which large farms may be less productive in terms of produce per surface area cultivated, but they own three quarters of the global agricultural land. The small-scale projects interviewed in this book and some of the others identified across Europe (see Appendix A), reduce the complexity of soil-less technology to make it affordable following a frugal innovation/alternative technology approach, with such simplified technology becoming operable by laypeople. Inevitably, in selling produce, these projects and enterprises have to compete in a market in which food is cheap and only large farms can produce sufficient quantities at a competitive price. If indoor urban farming expands, characteristics of the urban farming industry such as size, ownership, technology and business model will determine whether this industry will replicate the structure of the current agricultural system – with all its drawbacks – or if it will become one populated by a larger number of players, growing food according to shared values that are more in line with those underpinning food security and sovereignty rather than with predominantly financial returns.

9.4 Urban Agriculture and Soil-Less Technologies

The discussion on the shape of urban food production and the benefits that small rather than large farmers can generate, applies to a greater extent to in-soil urban agriculture generally, since this is the predominant form of food growing practiced in cities, while soil-less projects represent only a small fraction. But this fraction is nonetheless important because it shows that people in the urban agriculture community are receptive to innovation, helping them attain their aims. The opportunities offered by soil-less technologies are harnessed and adapted to the small scale, attracting the interest of members of local communities, armed with the skills and knowledge to use this technology. Within this process technology is internalised, debated and transformed. In being part of these projects, students (e.g., Real Food – see Sect. 6.3 - and Sow the City – see Sect. 6.12.1), urban dwellers (e.g., Poligono

Sur – see Sect. 6.1) and customers (e.g., Mangrovia Scicli – see Sect. 6.5 - and Bioaqua Farm- see Sect. 6.6) are made aware of and interact with a food technology which would be otherwise deployed far from them, in large production units detached from their lives. Technology is not understood in abstract terms but lived and experienced: it is not out of reach and control, but rather at hand. Concerns regarding the desirability, the degree to which it is artificial, and the safety of soil-less produce can be resolved through gaining direct experience of growing practices.

Can urban agriculture really transform the food system and, in doing so, bring issues such as the right to food and food poverty to the centre of the food systems debate? National and local plans for food security tend to focus more on the resilience of national supply chains, ignoring the fact that food is a highly politicised issue and that, even with sufficient food secured at a national level, access from disadvantaged groups to heathy food still needs to be ensured. The globalisation of food production and provision did not result in an alleviation of food poverty. The overall distribution of food is extremely unbalanced, with many industrialised countries producing and importing more food than necessary, resulting in a high share of food wasted. Yet food poverty is on the rise in both industrialised and industrialising countries. This suggests that food is conceived as a commodity rather than as a right to all. A more localised system of food production and distribution such as the urban agriculture model, can perhaps function as an alternative that can rectify the drawbacks of current centralised food chains.

For this to happen, urban agriculture needs to demonstrate that it can significantly contribute to food provision. As noted, much research has been concerned with modelling the potential for self-sufficiency of urban agriculture. Modelling was often based on space availability as the main factor rather than on the actual capabilities of farmers in terms of skills and organisation to maximise their production. Yet, some studies indicate that urban land and spaces can attain significant productivity. Not all soil-less urban farmers presented in the book attain high levels of production, compared to industrial benchmarks. Yet, they can be more productive than in-soil urban agriculture projects. A 2019 study in a London community garden recorded a yield of 1.34 kg/m^2 (Caputo et al., 2021). Other studies show yields that vary between 0.46 kg/m2 and 1.96 kg/m2 (Pourias et al., 2015), and between 1.99 and 15.53 kg/m2 (McDougall et al., 2019). Most of the case studies in this book go well beyond this in-soil productivity (e.g., Milagro de los Pisces 38.96 kg/m^2 and Mangrovia Scicli 23.57 kg/m^2 of crops harvested, without considering fish farming). More importantly, the food produced is likely to reach those who need it, thus improving the fairness of local food systems and probably contributing to reducing food waste.

9.5 Urban Soil-less Future Scenarios

Urban agriculture projects demonstrate that food is a political issue. Whether concerned with food security, healthy diets or the supply of zero-mile food, these projects move the focus from mere food supply to fairness, equity and quality of such supply. The difference between large scale, commercial and small-scale, community-led soil-less production is therefore one between the exploitation of a market gap within the broader industrial agriculture system and the exploration of a technology-enabled food system that is democratic, owned by all and therefore extremely diversified, and composed of small enterprises, backyard aquaponics, community-based social businesses and more. These projects form a patchwork of unique and distinct entities that produce more than food. They demonstrate the multi-dimensional character that food has in our culture, associating it with health, environmental awareness, social support or simply conviviality. These are projects that need to develop complex strategies to survive. They cannot rely on quantity of food produced and sold, for example, for lack of structure and capital. They can count on diversification of the offer and creativity. The exploration of new possibilities requires an open-minded view, flexibility and preparedness to change, with technology being just one of the factors that can help them achieve their goals. In their hands, technology is humanised. These are projects for which productivity needs to be redefined to go beyond food yields and include social gains. Is it unrealistic to think that current food production can be redesigned following an alternative model? Is the industrialised monoculture (of products and minds) the only possible option to feed the world? This is an ongoing debate, although the first step towards an alternative food system is to recognise that a radical change is possible.

One of the possible futures, based on a vision of soil-less urban farming, is promoted through studies on the industry in this sector. A representative of this industry is Plantagon, a Swedish R&D company with the mission of delivering 'sustainable, effective and esthetical 'agritechtural' solutions for governments and private enterprises globally' (Lauri, 2017). Plantagon has developed soil-less indoor systems, registered many patents, provided consultancy to many indoor vertical farms worldwide and applied for planning permission for a large-scale demonstrator of their technologies in the municipality of Linköping, 200 km from Stockholm. The demonstrator is called the World Food Building and deploys technologies that they designed such as a helix conveying system and a heat turbulence control system which will enable higher resource efficiency. The demonstrator also promotes a vision of high-tech indoor urban farming with an architectural language that transpires modernity and progress. In fact, the initial project was designed within a futuristic, iconic, large scale, transparent, double-skin geodesic dome. The final design, changed after opposition of local groups on the basis that the transparent dome could be dangerous for birds, is a tower engineered to grow between 300,000 and 500,000 kg of vegetables per year on the south façade (Lauri, 2017). Thought-provoking and fascinating, Plantagon's proposition is designed to attract big investors. The underlying, although not explicitly stated, assumption is that urban soil-less produce is sustainable, and that increasing its production equates to reduced environmental impact and food availability for a growing population. This narrative, as noted, has some drawbacks and does not consider the dynamics of food poverty which negate access of low-income groups to healthy food; the political dimension of food is excluded from this vision.

Most studies on vertical farming reference Despommier as one of the eminent promoters of this concept. One of his main arguments is that cities can offer space (waste land and buildings), expertise (as places with high concentration of knowledge), resources (waste heat and water) and demand that make high-tech food production economically viable and environmentally beneficial. Through careful design, vertical farms can be powered by a sufficiently large surface of PV panels (Al-Chalabi, 2015). In some studies, vertical farming is recognised as an urban type of food production which can lead to social benefits too, such as educational, psychological and leisure related (Kalantari et al., 2018). High-tech urban farming cannot be a substitute for agriculture but as a sector can grow alongside in-soil production. Vertical farming, in particular, is enthusiastically promoted as the most appropriate food technology for an urban context, thus requiring further in-depth research and investment to make it more viable (Beacham et al., 2019). To date, however, there are only a few companies that use this technology on a medium scale. Birkby (2016) provides a list of 14 companies operating in 2015 in the USA, predicting that the number is expected to treble by the end of 2016.

In Birkby's list, the majority of these companies is either a technology provider (e.g., small scale (CropBox[1] – delivering shipping-container vertical farming systems) or is small (e.g., Local Garden[2] - two-storey, 560 m^2 building on the roof of a parking garage). The largest in the list is Aerofarms, a company in New Jersey, USA, which has established many farms, with the most important one (apparently the largest vertical aeroponic farm worldwide) located in a former 7000m^2 steel mill, producing a two million pounds (900,000 kg) harvest per year (Aerofarms, 2021). Other large controlled environment agriculture companies include Plenty, Brightgrow, Zipgrow, Spread Co. and Sanan Sino-Science Photobiotech Co. Innovatus, a Japanese company, grows 12,000 heads of lettuce per day within an approximately 2000 m^2 facility (O'Sullivan et al., 2019; Skyfarn, 2021). Kalantari et al. (2018) also mention Nuvege Plant Factory in Kyoto, Japan (about 3000m^2 surface area) and Green Sense Farm in Shenzhen, China (about 2000 m^2 surface area). We are very far from the critical mass needed to implement the vision of a city in which high-tech food enables local self-sufficiency.

Vertical Farming seems to be a model capable of capturing the imagination of scholars, researchers and architects, but more 'conventional' soil-less farms, with a more modest appearance, are also being developed. In parallel, consultancies – perhaps less ambitious than Plantagon - advising on technology, architecture and engineering, and business models are now operating. Agritecture, for example, is a consultancy company based in New York, that coined their name as a term capturing 'the art, science, and business of integrating agriculture into cities'. They have designed vertical farms such as Farm One in New York and Dream Harvest in Houston, as well as hydroponic systems on rooftops such as Sky Vegetables, an 800m^2 rooftop farm in New York, on an office building. Agripolis, a consultancy

[1] www.cropbox.co

[2] www.localgarden.com

company in Paris, designs and installs hydroponic/aeroponic farms, particularly on rooftops and using towers. They have developed several projects for supermarkets such as Carrefour (S.te Genevieve des Bois) and companies (Issy les Moulineaux - Sodexo World headquarters), on rooftops and open spaces, organising crop production outdoors. Other companies are developing small hydroponic units that can be used at home or even in retailer outlets. NeoFrames is a German company proposing top-end hydroponic cabinets that can be integrated into kitchen systems. Infarm is developing and marketing hydroponic units to grow microgreens directly in supermarkets. Beyond urban farming, R&D into soil-less is receiving great attention, especially in the field of IT applied to the management of hydroponic/aquaponic systems, developing 'smart' solutions. On a smart soil-less agriculture farm, for example, sensors can read the composition of the nutrient solution, analyse it and correct it when necessary, with great improvements in productivity (Sambo et al., 2019).

It is difficult to predict whether the future of urban soil-less farming is in the large-scale futuristic vertical farm city, a vision chiming with that of smart cities, or in a more diffused web of small farms, composed of backyard aquaponic units, community-led projects and small enterprises, which recognise the right to food as a tenet of food systems. In this book, case studies presented as well as the analysis of their productive capacity and values driving the efforts of soil-less farmers, show an uncommon energy and willingness to integrate new food technologies within their projects. In Chap. 2, the propensity within urban agriculture to experiment with new arrangements and – in the case of soil-less projects – technologies, has been pointed out as one of the characteristics of a practice that has evolved significantly over the last few decades. In many cities, the urban agriculture scene relies on organised networks of food growers, exchanging information, promoting events and debating policies. These networks function as living labs, with projects testing new models of social support to low-income groups, minority groups, or even elderly people affected by dementia (Noone & Jenkins, 2018; Ghose & Pettygrove, 2014; Yotti' Kingsley & Townsend, 2006), new models of food distribution based on – for example - CSA (Community Supported Agriculture, 2021), and creative approaches to ensure financial viability such as those developed by RotterZwam (see Sect. 6.10) and Trueffelwerk (see Sect. 6.11). Urban agriculture is functioning as a 'space' in which diverse agendas – driven from below – converge through the action of food growing. Food is not only vital but also a cornerstone of human culture, with social and political implications.

Yet most of the farms interviewed as well as those identified during the investigation for this book produce mainly leafy greens as these crops can generate higher profits, which is a necessary condition for their economic viability. This very narrow range of crops may result in a fast saturation of the market, in a scenario of higher uptake of soil-less technologies. Potentially, whether vertical or horizontal, hydroponics can produce a large variety of crops, although it seems sensible to grow staple crops on sufficiently large land areas, rather than in cities where land is limited. Cereals, potatoes and high calorie crops, necessitate vast cultivation areas to meet demand. O'Sullivan et al. (2019) advocate for advancements in breeding crops with traits that are suitable for indoor vertical farming. Traits such as resistance to drought and cold temperatures are unnecessary for soil-less production while, for example, different canopy architectures could make crops more suitable for this technology. It would also be necessary for many fruit plants to adapt to lower light conditions. Research in this direction can help expand the current range of soil-less urban produce and increase its contribution to meet demand for a larger variety of crops.

9.6 Unnatural

Food technologies can evoke artificiality and a sense of unnatural, which are not typically associated with healthy food, and can therefore be met with diffidence. Recent development of *Public Understanding of Science* suggests that science must be understood, trusted and 'owned' by people in order to be accepted. Imposition of technological progress risks rejection. Urban farmers have demonstrated great interest in the technical aspects of horticulture, just as other have experimented with organic techniques (Garg & Balodi, 2014) and agroecology principles (Altieri & Nicholls, 2020). The case studies in this book demonstrate that urban farmers can adopt and integrate technology in their projects. These case studies represent only a small niche within the broader networks of projects, but an analysis of Google searches and the identification of YouTubers dedicated to soil-less cultivation suggest that the phenomenon is not so small and growing. Appendix A shows the small to medium projects identified during the investigation for this book. The table includes not only urban gardens and farms but also organisations, experimental projects and consultancy companies, demonstrating that soil-less projects attract the attention of many diverse groups, and it is therefore already a phenomenon with many ramifications.

One of the questions posed in the introduction to the book is whether the interaction with food technology is a sign that the relationship between urban farmers and nature is changing. In accepting that soil-less technology does not equate to artificial and unnatural, and that synthetic nutrients used in this process do not damage people and soil health, are urban farmers moving away from a vision of urban agriculture as a process of ecological amelioration in cities? The case studies suggest that soil-less urban farmers either perceive this technology as one of the possible ways to grow food sustainably, one of the latest additions to a long tradition of horticultural techniques that, in cities more than in rural areas, help bring people close to and understand nature; or the adoption of soil-less is accepted on the basis that nature has been damaged to the point that it can no longer be trusted. Urban farmers using aquaponic units as demonstrators of plant and fish physiology, view such a technology as a tool or an 'interactive textbook' showing how nature works. In these projects, crop and fish growth is 'decontextualised' and displayed out of the natural media and habitat of plants and animals (e.g., roots in transparent containers and fish in tanks), although not using sophisticated equipment and digitalised control systems but rather a DIY approach that perhaps makes the perception of this technology less artificial. In this way, visitors are not interacting with machines to grow crops but with tools, hardly noticeable, colonised by plant growth. This setting mitigates the conceptual implications of a food production that is under the control of specialists and specialist machines. Alternatively, in the case of projects such as Mangrovia Scicli and BioAqua Farm, the main rationale for using these food production technologies is the general loss of trust in industrial agriculture and farming. These industries operate with methods that have damaged nature and produce unsafe food. Soil-less technology can rectify this by reducing synthetic inputs and controlling quality of water and growing media. Enterprises such as GrowUp share the same concerns on the quality of food and concentrate their efforts on reducing the footprint of food production. For these farmers, soil-less technologies are an opportunity to improve global natural conditions rather than mainly local ones.

The border between artificial and natural is blurred. Technology can be used to control and transform nature but also to enhance it and, in the process, perhaps interfere with it. Precision agriculture offers powerful tools to rationalise resource use and is perhaps a step forward for the agricultural intensification advocated by FAO. But how this will help reduce demand for agricultural land, which is also determined by diets and the market rather than solely demographic growth, is less clear. Technology deployed without political purposes can be damaging. Urban farmers have powerful visions and purpose for soil-less technologies, ensuring that their appreciation of nature is not hindered by the technological artificiality of such technologies.

9.7 Small and Large, Urban Waste and Final Remarks

In a future scenario in which small scale soil-less urban agriculture has a stronger presence in cities, how do we distinguish between large and small? Can the scale of production influence the values that drive social and for-profit enterprises, which can support a more equitable food system? While these questions cannot be answered within the context of this book, a few reflections can contribute to the debate. At a global perspective, an agricultural industry in which about 16% of farms cultivate 88% of worldwide agricultural land (Lowder et al., 2016) is particularly vulnerable; the failure of a small number of these farms for environmental, political or financial reasons, can severely disrupt the global supply chain. Redundancy is an attribute of resilience (Peterson et al., 1998). Increasing the number of farms should decrease risks of disruption in systems. Large farms employing a high number of workers must also guarantee a cashflow, and this, in turn, can influence decisions on the sustainability of practices used for cultivation, when these are deemed to negatively affect levels of productivity and financial returns. Instead, small enterprises are nimbler and faster in reacting to changes and redirecting their practices when necessary. But to assume that the scale of the farms alone can reduce the impact on the environmental footprint of food and increase the resilience of global food systems would be an over-simplification and a mistake. Policy could correct many of these potential drawbacks with an effective system of incentives and tighter environmental standards of food production. Perhaps what should be noted is that the urban context, generally, can restrict the scale of intervention of urban agriculture, and the only food production model that can be implemented in cities is one that relies on small-to-medium production. Soil-less systems have the advantage of being highly productive when productivity is measured as a unit of output on a unit of surface area of production. Small hydroponic farms in terms of surface area can compare with medium conventional farms in terms of yield, which again, makes them particularly suitable for cities. Urban agriculture can also become a key factor in a scenario of increased national food security, which some authors advocate as necessary in light of climate change and the implementation of local political strategies that may disrupt the globalised supply chain, such as Brexit (Lang, 2020a, 2020b).

It should be also recognised that by operating in an urban context, these enterprises and projects establish a tighter relationship with people, producing social benefits that are linked to food production. It would be difficult for large scale farms (and rural farms) to achieve this level of engagement. Food justice (which focuses on the ethical dimension of food systems, promoting fair access to food for all – Lang, 2020a, 2020b) needs to be embedded in a sustainable food system, since higher productivity alone will not alleviate food poverty. Studies suggest that urban agriculture has the potential to mitigate food injustice (Santo et al., 2016; Tornaghi, 2017) although for this to happen it is necessary for groups experiencing food insecurity to directly benefit from urban agriculture (Horst et al., 2017; Meenar & Hoover, 2012). The lack of professional experience in horticultural production of many volunteers working in community gardens and farms as well as the prioritisation of social benefits connected with farming may result in relatively low productivity (Kirby et al., 2021). Lower productivity may be counterbalanced with shorter supply chains and higher returns to urban farmers. An organic standard extended to soil-less produce could help in increasing its uptake because of the added value that customers are prepared to pay for organic food. But this may not be sufficient and the aid of policy in recognising through incentives the significant contribution to social amelioration that urban agriculture projects provide may be needed.

Cities generate waste, which can be turned into resources. Food waste can become compost, rainwater can be harvested, and waste heat can be harnessed for greenhouses. The rooftop greenhouse experiment in Berlin goes one step further and demonstrates that wastewater too can be utilised for the production of healthy food. A small enterprise in London is experimenting with the by-product of anaerobic digestion as a nutrient in soil-less production (LEAP, 2020). Researchers are investigating affordable and operationally manageable ways to produce phosphorous from fish waste (Goddek et al., 2015). Redesigning the metabolism of cities by redirecting waste streams into food production can be a great leap towards sustainable cities. But the scale of modification of urban systems necessary for this to happen is massive: the reorganisation of waste collection, and in particular the transfer of waste heat to indoor farm units are perhaps highly complex tasks to implement.

They are nevertheless great opportunities that are attracting the focus of researchers and should inform research programmes to a greater extent.

A concluding remark reiterates the claim underpinning the need for the book that, although marginal, the phenomenon of small-scale soil-less will grow and therefore should be analysed. The projects listed in Appendix A as well as the interest catalysed by soil-less farming on the web, documented in Chap. 7, show that the phenomenon is growing. Food is a topic that is increasingly attracting concern and attention. In addition, the extreme uncertainty about the direction towards which macro drivers of change will direct the not-too-distant future of the planet seem to point at the need to increase levels of local self-sufficiency. While urban agriculture has this strategy at its core, technology, when used in an environmentally and socially sustainability way, can augment it. The case studies in this book offer different approaches to integrate technology into an urban food production system that are sometimes equitable and always innovative. If anything, the book has documented these approaches and their productive and environmental characteristics. As such, it can be used as a compendium of future pathways for urban agriculture.

Correction to: The Web Community of Soil-Less Farmers: A Case Study

Valentina Manente and Silvio Caputo

Correction to:
Chapter 7 in: S. Caputo, *Small Scale Soil-less Urban Agriculture in Europe*, Urban Agriculture, https://doi.org/10.1007/978-3-030-99962-9_7

The original version of the chapter was inadvertently published without adding author's name "Valentina Manente" in chapter 7. The corrected version includes the name of the author. The authors of chapter 7 are Valentina Manente and Silvio Caputo.

The updated version of the chapter can be found at https://doi.org/10.1007/978-3-030-99962-9_7

S. Caputo, *Small Scale Soil-less Urban Agriculture in Europe*, Urban Agriculture, https://doi.org/10.1007/978-3-030-99962-9_10

Appendixes

Appendix A

The projects included in this table were active at the completion of the book, in April 2021.

Hydroponic projects and farms – SMALL	Aquaponic projects and farms – SMALL	Mushrooms projects and farms
Crop Cycle, Green Meadow Community Farm in Cwmbran & Treherbert, Wales, UK – https://businesswales.gov.wales/foodanddrink/news-and-events/news/crop-cycle-welsh-government-fund-supply-installation-controlled-environment-agriculture-systems	Agriloops, Paris. France – https://www.agriloops.com/	Containing mushrooms, Hoofddorp, Netherlands – https://containingmushrooms.nl/
Fives Cail, Lille, France – https://www.cueilletteurbaine.com/module-agricole-fives-cail/	BioAqua Farm, Wedmore, UK – www.bioaquafarm.co.uk	GrowCycle Exeter https://grocycle.com/urban-mushroom-farm/
Fridheimar, Selfoss, Iceland – https://www.fridheimar.is/en	Bristol Fish Project. Bristol, UK – https://bristolfish.org/	Helsieni, Finland – https://www.helsieni.fi/en/home/
Glasgow Greens, UK – https://www.glasgowgreens.co.uk/	De Ceuvel, Amsterdam, Netherlands – http://deceuvel.nl/en/about/sustainable-technology/	La Caverne, Paris, France https://lacaverne.co/en/cavern-urban-farm/
Growing Underground, London, UK – http://growing-underground.com/	El Milagro de los Peces, Seville, Spain – http://www.iesromeromurube.es/	RotterZwam, Rotterdam, Netherlands – http://rotterzwam.nl

(continued)

S. Caputo, *Small Scale Soil-less Urban Agriculture in Europe*, Urban Agriculture, https://doi.org/10.1007/978-3-030-99962-9

Hydroponic projects and farms – SMALL	Aquaponic projects and farms – SMALL	Mushrooms projects and farms
GROWx, Amsterdam, Netherlands – https://www.growx.co/about	Huerto Lazo, Cajiz, Spain – www.huertolazo.eu	Trueffelwerk, Recklinghauser, Germany – https://www.trueffelwerk-recklinghausen.de/
Grow Bristol, UK – http://growbristol.co.uk	Espace BSA, Paris, France – https://www.cueilletteurbaine.com/laquaponie-dans-tous-ses-etats/	Hut & Stiel, Vienna, Austria–- https://www.hutundstiel.at/
Kantapoya, Helsinki, Finland – http://kaantopoyta.fi/	Fesch Haff, Machtum, Luxembourg – https://www.fesch-haff.lu/	Nãm, Lisbon, Portugal – https://nammushroom.com/
Space Farms, Tbilisi, Georgia – https://spacefarms.ge/	Mangrovia Scicli, Scicli, Italy – www.mangroviaproject.com	Funga Farm, Copenhagen, Denmark – https://yeswefood.com/fungafarm
Vertigo Farms, Puławy, Poland – https://vertigofarms.eu/o-nas/	UGH, Dortmund, Germany – www.derubanisten.de/https://dieurbanisten.de/aquaponikgewaechshausfuehrungen/	Dirfis Mushrooms, Euboea, Greece–- https://manitariadirfis.gr/index.php
Hydroponic projects and farms – MEDIUM	Urban Smart Farm, Ghent, Belgium http://www.urbansmartfarm.be/en/	
Cite Maraichare, Romanville, France https://www.lacitemaraichere.com/en/	Real Food Wythenshawe, Wythenshawe, UK – https://www.realfoodwythenshawe.com	
Farma Rajecek, Brno, Czech Republic – https://farmarajecek.cz/cz	**Aquaponic projects and farms – MEDIUM**	
Ste Genevieve des Bois - Hypermarket (Agripolis), Paris, France – http://agripolis.eu/project/ste-genevieve-des-bois-hypermarche/	Blue Acres. Eindhoven, Netherlands – http://www.blueacres.nl/ (http://www.hortidaily.com/article/34932/Dutch-entrepreneur-receives-award-for-aquaponics-technology)	
Green Jungle, Portugal – https://greenjungle.pt/	Duurzamekost, Eindhoven, Netherlands https://duurzamekost.nl/	
Happy Fruits, Novo Selo, Bulgaria – http://happyfruits.eu/en/pages/greenhouse.php	Ecco-Jaeger, Switzerland – http://www.ecco-jaeger.ch/	
Issy les Moulineaux – Sodexo World headquarters, France – http://agripolis.eu/project/issy-sodexo/	Duurzame Kost, Eindhoven, Netherlands – http://www.duurzamekost.nl/	
Ljusgarda, Tibro, Sweden https://ljusgarda.se/	Farminova, Kisim, Turkey – https://www.farminova.com/	

(continued)

Hydroponic projects and farms – SMALL	Aquaponic projects and farms – SMALL	Mushrooms projects and farms
Planet Farms, Italy – https://www.planetfarms.ag/en/our-world	Flenexa Aquaponie, Přáslavice, Czech Republic – https://www.aquaponickafarma.cz/en/home-aquaponick-farm-praslavice/	
Urban Oasis, Stockholm, Sweden https://www.urbanoasisfarming.com/	GrowUp – Unit 84, London, UK – https://www.growupfarms.co.uk/our-history/	
Valgio' Orti Sociali, Rome. Italy – http://www.valgio.it/gli-orti-sociali/	Stadfarm, Berlin, Germany – https://www.stadtfarm.de/en/	
Veles Farming, Slovakia – https://www.velesfarming.com/home-english#benefits	**Educational aquaponic projects**	
Educational hydroponic projects	Bio T Fuel, Nantes, France – https://bio-t-full.org/	
Hemmaodlat, Malmo, Sweden – https://www.hemmaodlat.se/	Mediamatic, Amsterdam, Netherlands – https://www.mediamatic.net/en/page/74034/aquaponics	
Sow the City, Manchester, UK – https://www.sowthecity.org/	**Experimental-research projects (*) and technology providers (**)**	
Experimental projects (*) and technology providers (**)	Aquapioners**, Barcelona, Spain-http://aquapioneers.io/	
FarmUrban**, UK – www.farmurban.co.uk	Aquaponics Iberia, Turcifal – Torres Vedras,Portugal – https://www.aquaponicsiberia.com/	
Grönska**, Stockholm, Sweden – https://www.gronska.org/	Green Lab*, London, UK – https://www.greenlab.org/#facilities	
Infarm**, Berlin, Germany – https://infarm.de/	PGC**, Belgium – http://www.pcgroenteteelt.be/nl-nl/	
Impact Farm*, Copenhagen, Denmark http://www.humanhabitat.dk/project-1	Roof Water Farm*, Berlin, Germany – http://www.roofwaterfarm.com/en/ueber/	
LettUs Grow, Bristol, UK** – https://www.lettusgrow.com/		
Micro Flavours Brussels, Molenbeek, Belgium – https://microflavours.brussels/#top		
NeoFarms**, Germany – https://www.neofarms.com/		

(continued)

Hydroponic projects and farms – SMALL	Aquaponic projects and farms – SMALL	Mushrooms projects and farms
R&D facility ReFarmers*, Lyon, France – https://refarmers.co/about/pilot-farm/		
Urban Crop Solutions**, Belgium https://urbancropsolutions.com/modulex/		
Vertical Farm Lab**, Amsterdam, Netherlands - https://www.onefarm.io/		
Vertical Future**, London, UK – https://www.verticalfuture.co.uk/		

Appendix B

List of Youtubers included in the sample analysed in Chap. 7 , with the link to one of their most representative videos reviewed in the study.

Youtuber	Video link	Views on 01/12/2020	Date posted
1	https://www.youtube.com/watch?v=FKd-SrkqEfU&list=PLf3JjDpi4FW75WLagKd01YoCpJ0tcMHPF	6687 views	2019/06/27
2	https://www.youtube.com/watch?v=JXKfIASdSqM&list=PLf3JjDpi4FW75WLagKd01YoCpJ0tcMHPF&index=3&t=6s	950,353 views	2018/04/13
3	https://www.youtube.com/watch?v=3AbxnWiVTH8&list=PLf3JjDpi4FW75WLagKd01YoCpJ0tcMHPF&index=3	530,931 views	2019/11/29
4	https://www.youtube.com/watch?v=A-DNrQoKcnc&list=PLf3JjDpi4FW75WLagKd01YoCpJ0tcMHPF&index=5&t=0s	1,082,784 views	2015/03/08
5	https://www.youtube.com/watch?v=-VFwKDZQxAg&list=PLf3JjDpi4FW75WLagKd01YoCpJ0tcMHPF&index=6	432,875 views	2017/10/30
6	https://www.youtube.com/watch?v=K7Fho-kphMA&list=PLf3JjDpi4FW75WLagKd01YoCpJ0tcMHPF&index=11&t=199s	433,706 views	2017/05/21
7	https://www.youtube.com/watch?v=VhsCcR7YEXI&list=PLf3JjDpi4FW75WLagKd01YoCpJ0tcMHPF&index=16	95,368 views	2015/02/09
8	https://www.youtube.com/watch?v=PqArMDSq4Fo&list=PLf3JjDpi4FW75WLagKd01YoCpJ0tcMHPF&index=18	220,355 views	n/a
9	https://www.youtube.com/watch?v=nXy32Dr4Z4A&list=PLf3JjDpi4FW75WLagKd01YoCpJ0tcMHPF&index=24	2,984,677 views	2012/10/08
10	https://www.youtube.com/watch?v=jGMOaczYhOg&list=PLf3JjDpi4FW75WLagKd01YoCpJ0tcMHPF&index=25	381,825 views	2017/05/25
11	https://www.youtube.com/watch?v=0dyoIiNyvNI&list=PLf3JjDpi4FW75WLagKd01YoCpJ0tcMHPF&index=29&t=20s	150,443 views	2018/03/20

(continued)

Youtuber	Video link	Views on 01/12/2020	Date posted
12	https://www.youtube.com/watch?v=LZk2-y74SVw&list=PLf3JjDpi4FW75WLagKd01YoCpJ0tcMHPF&index=31&t=170s	167,672 views	2018/11/28
13	https://www.youtube.com/watch?v=12QEwDROvzk&list=PLf3JjDpi4FW75WLagKd01YoCpJ0tcMHPF&index=36&t=146s	22,284 views	2020/01/14
14	https://www.youtube.com/watch?v=t2RDmC1JHXk&list=PLf3JjDpi4FW75WLagKd01YoCpJ0tcMHPF&index=40	681,532 views	2016/03/04
15	https://www.youtube.com/watch?v=0NVxrq2SKxs&list=PLf3JjDpi4FW75WLagKd01YoCpJ0tcMHPF&index=45	716,169 views	2016/03/27
16	https://www.youtube.com/watch?v=M9BBbU87vUw&list=PLf3JjDpi4FW75WLagKd01YoCpJ0tcMHPF&index=51&t=21s	14,603 views	2019/12/20
17	https://www.youtube.com/watch?v=TxNSQd_OQMo&list=PLf3JjDpi4FW75WLagKd01YoCpJ0tcMHPF&index=70	8489 views	2013/10/04
18	https://www.youtube.com/watch?v=1PVdQ3-7UIs&list=PLUoIqx1-oCwj3RrKBb8xmXmwW8Gf9k_no	866,719 views	2017/06/23
19	https://www.youtube.com/watch?v=paEbCyNH2Ok&list=PLf3JjDpi4FW5GcI1SUm86XU1xO0_Fn_Po&index=3&t=0s	91,474 views	2019/12/15
20	https://www.youtube.com/watch?v=k5HjG6yYB1s&list=PLf3JjDpi4FW5GcI1SUm86XU1xO0_Fn_Po&index=7&t=0s	345,520 views	2012/10/21
21	https://www.youtube.com/watch?v=6vxuO0PQzws&list=PLf3JjDpi4FW5GcI1SUm86XU1xO0_Fn_Po&index=10&t=249s	44,789 views	2016/05/30
22	https://www.youtube.com/watch?v=M21GmwmPq1w&list=PLf3JjDpi4FW5GcI1SUm86XU1xO0_Fn_Po&index=16&t=567s	124,953 views	2018/10/26
23	https://www.youtube.com/watch?v=o8n8h2ljrP8&list=PLf3JjDpi4FW5GcI1SUm86XU1xO0_Fn_Po&index=12&t=0s	32,474 views	2013/10/09
24	https://www.youtube.com/watch?v=IVSEmwY2u0I&list=PLf3JjDpi4FW5GcI1SUm86XU1xO0_Fn_Po&index=14	1,414,759 views	2014/10/11
25	https://www.youtube.com/watch?v=tKNuoYCdue4&list=PLf3JjDpi4FW5GcI1SUm86XU1xO0_Fn_Po&index=17&t=894s	497,670 views	2014/01/01

(continued)

Youtuber	Video link	Views on 01/12/2020	Date posted
26	https://www.youtube.com/watch?v=WGI6O8FqLtA&list=PLf3JjDpi4FW5GcI1SUm86XU 1xO0_Fn_Po&index=20	784,261 views	2014/11/23
27	https://www.youtube.com/watch?v=pUa_UixXRfk&list=PLf3JjDpi4FW5GcI1SUm86XU 1xO0_Fn_Po&index=35&t=36s	878,901 views	2014/08/25
28	https://www.youtube.com/watch?v=G5w-6pnjwIc&list=PLf3JjDpi4FW5GcI1SUm86XU 1xO0_Fn_Po&index=41	225,128 views	2014/06/01
29	https://www.youtube.com/watch?v=YZKGg7mHjPk&list=PLf3JjDpi4FW5GcI1SUm86XU 1xO0_Fn_Po&index=43	310,467 views	2015/04/19
30	https://www.youtube.com/watch?v=0zCJFolqLY4&list=PLf3JjDpi4FW5GcI1SUm86XU 1xO0_Fn_Po&index=84	69,649 views	2012/10/17

References

Ackerman, K., Conard, M., Culligan, P., Plunz, R., Sutto, M. P., & Whittinghill, L. (2014). Sustainable food systems for future cities: The potential of urban agriculture. *The Economic and Social Review, 45*(2), 189–206.

Acton, L. (2011). Allotment gardens: A reflection of history, heritage, community and self. *Papers from the Institute of Archaeology*, 21.

Acton, L. (2015). *Growing space: A history of the allotment movement*. Five Leaves Publications.

Adams, D., & Hardman, M. (2014). Observing guerrillas in the wild: Reinterpreting practices of urban guerrilla gardening. *Urban Studies, 51*(6), 1103e1119.

Adamsone-Fiskovica, A. (2015). Technoscientific futures: Public framing of science. *Technology in Society, 40*, 43–52.

Adeoti, A. I., Cofie, O., & Oladele, O. I. (2012). Gender analysis of the contribution of urban agriculture to sustainable livelihoods in Accra, Ghana. *Journal of Sustainable Agriculture, 36*(2), 236–248.

Agrylist. (2016). *State of indoor farming*. Agrylist. Available at on https://artemisag.com/state-of-indoor-farming-2016/. Accessed 10.09.2020.

Aker, J. C., & Mbiti, I. M. (2010). Mobile phones and economic development in Africa. *Journal of Economic Perspectives, 24*(3), 207–232.

Al-Chalabi, M. (2015). Vertical farming: Skyscraper sustainability? *Sustainable Cities and Society, 18*, 74–77.

Altieri, M. A., & Nicholls, C. (2020). Agroecology: Challenges and opportunities for farming in the Anthropocene. *International Journal of Agriculture and Natural Resources, 47*(3), 204–215.

Altieri, M. A., Nicholls, C. I., Henao, A., & Lana, M. A. (2015). Agroecology and the design of climate change-resilient farming systems. *Agronomy for Sustainable Development, 35*(3), 869–890.

Andersson, E., Barthel, S., Borgström, S., Colding, J., Elmqvist, T., Folke, C., & Gren, A. (2014). Reconnecting cities to the biosphere: Stewardship of green infrastructure and urban ecosystem services. *Ambio*. https://doi.org/10.1007/s13280-014-0506-y

Arnon, D. I., & Hoagland, D. R. (1944). The investigation of plant nutrition by artificial culture methods. *Biological Reviews, 19*(2), 55–67.

Aschemann-Witzel, J., Gantriis, R. F., Fraga, P., & Perez-Cueto, F. J. (2020). Plant-based food and protein trend from a business perspective: Markets, consumers, and the challenges and opportunities in the future. *Critical Reviews in Food Science and Nutrition*, 1–10.

Atkinson, N. (2018). Hydroponics: Should we think small? *CPQ Nutrition, 1*(3), 1–17.

Aubry, C., & Daniel, A. C. (2017). Innovative commercial urban agriculture in the Paris metropolitan area. In *Toward sustainable relations between agriculture and the city* (pp. 147–162). Springer.

S. Caputo, *Small Scale Soil-less Urban Agriculture in Europe*, Urban Agriculture, https://doi.org/10.1007/978-3-030-99962-9

Aulie, R. P. (1970). Boussingault and the nitrogen cycle. *Proceedings of the American Philosophical Society, 114*(6), 435–479.

Ausubel, J. H., Wernick, I. K., & Waggoner, P. E. (2013). Peak farmland and the prospect for land sparing. *Population and Development Review, 38*, 221–242.

Bacon, C. M., & Baker, G. A. (2017). The rise of food banks and the challenge of matching food assistance with potential need: Towards a spatially specific, rapid assessment approach. *Agriculture and Human Values, 34*(4), 899–919.

Badami, M. G., & Ramankutty, N. (2015). Urban agriculture and food security: A critique based on an assessment of urban land constraints. *Global Food Security*. https://doi.org/10.1016/j.gfs.2014.10.003

Baddadi, S., Bouadila, S., Ghorbel, W., & Guizani, A. (2019). Autonomous greenhouse microclimate through hydroponic design and refurbished thermal energy by phase change material. *Journal of Cleaner Production, 211*, 360–379.

Bailey, D. S., & Ferrarezi, R. S. (2017). Valuation of vegetable crops produced in the UVI commercial Aquaponic system. *Aquaculture Reports, 7*, 77–82.

Barbosa, G. L., Gadelha, F. D. A., Kublik, N., Proctor, A., Reichelm, L., Weissinger, E., Wohlleb, G. M., & Halden, R. U. (2015). Comparison of land, water, and energy requirements of lettuce grown using hydroponic vs. conventional agricultural methods. *International Journal of Environmental Research and Public Health, 12*(6), 6879–6891.

Barrett, G. E., Alexander, P. D., Robinson, J. S., & Bragg, N. C. (2016). Achieving environmentally sustainable growing media for soilless plant cultivation systems–a review. *Scientia Horticulturae, 212*, 220–234.

Barthel, S., Folke, C., & Colding, J. (2010). Social-ecological memory in urban gardens - retaining the capacity for management of ecosystem services. *Global Environmental Change, 20*(2), 255e265.

Barthel, S., Parker, J., & Ernstson, H. (2013). Food and green space in cities: A resilience lens on gardens and urban environmental movements. *Urban Studies, 52*(7), 1321–1338.

Barthel, S., Parker, J., & Ernstson, H. (2015). Food and green space in cities: A resilience lens on gardens and urban environmental movements. *Urban Studies, 52*(7), 1321–1338.

Bauer, M. W., Allum, N., & Miller, S. (2007). What can we learn from 25 years of PUS survey research? Liberating and expanding the agenda. *Public Understanding of Science, 16*(1), 79–95.

Baur, J. (2020). Campus community gardens and student health: A case study of a campus garden and student Well-being. *Journal of American College Health*, 1–8.

Beacham, A. M., Vickers, L. H., & Monaghan, J. M. (2019). Vertical farming: A summary of approaches to growing skywards. *The Journal of Horticultural Science and Biotechnology, 94*(3), 277–283.

Beebe, J. K., Amshoff, Y., Ho-Lastimosa, I., Moayedi, G., Bradley, A. L., Kim, I. N., Casson, N., Protzman, R., Espiritu, D., Spencer, M. S., & Chung-Do, J. J. (2020). Reconnecting rural native Hawaiian families to food through aquaponics. *Genealogy, 4*(1), 9.

Bell, E., Etzel, R., Hammer, B., Frey, L., Harmon, T., Blank, T., Meeusen, C., Burn, B., Schon, M., Huang, Y., & Petitt, F. (2004). *Food crop culture in the land greenhouses at EPCOT®*. In VII International symposium on protected cultivation in mild winter climates: production, pest management and global competition 659 (pp. 161–169).

Bell, S., Fox-Kämper, R., Keshavarz, N., Benson, M., Caputo, S., Noori, S., & Voigt, A. (Eds.). (2016). *Urban allotment gardens in Europe*. Routledge.

Bellarby, J., Tirado, R., Leip, A., Weiss, F., Lesschen, J. P., & Smith, P. (2013). Livestock greenhouse gas emissions and mitigation potential in Europe. *Global Change Biology, 19*(1), 3–18.

Benis, K., Gomes, R., Vicente, R., Ferrao, P., & Fernandez, J. (2015). Rooftop greenhouses: LCA and energy simulation. In Proceedings of International conference CISBAT 2015 future buildings and districts sustainability from nano to urban scale (No. conf, pp. 95–100). LESO-PB, EPFL.

Berardi, U., GhaffarianHoseini, A., & GhaffarianHoseini, A. (2014). State-of-the-art analysis of the environmental benefits of green roofs. *Applied Energy, 115*, 411–428.

Bergek, A., Jacobsson, S., Carlsson, B., Lindmark, S., & Rickne, A. (2008). Analyzing the functional dynamics of technological innovation systems: A scheme of analysis. *Research Policy, 37*, 407–429.

Berners-Lee, M., Kennelly, C., Watson, R., Hewitt, C. N., Kapuscinski, A. R., Locke, K. A., & Peters, C. J. (2018). Current global food production is sufficient to meet human nutritional needs in 2050 provided there is radical societal adaptation. *Elementa: Science of the Anthropocene, 6*.

Birkby, J. (2016). Vertical farming. *ATTRA Sustainable Agriculture*, 1–12.

Blue, G. (2010). Food, publics, science. *Public Understanding of Science, 19*(2), 147–154.

Bolund, P., & Hunhammar, S. (1999). Ecosystem services in urban areas. *Ecological Economics, 29*(2), 293–301.

Boneta, A., Rufí-Salís, M., Ercilla-Montserrat, M., Gabarrell, X., & Rieradevall, J. (2019). Agronomic and environmental assessment of a polyculture rooftop soilless urban home garden in a Mediterranean city. *Frontiers in Plant Science, 10*, 341.

Borch, A., & Kjærnes, U. (2016). Food security and food insecurity in Europe: An analysis of the academic discourse (1975-2013). *Appetite*. https://doi.org/10.1016/j.appet.2016.04.005

Bostan, I., Onofrei, M., Toderaşcu, C., & Lazăr, C. M. (2019). An integrated approach to current trends in organic food in the EU. *Food, 8*(5), 144.

Bramley, R. G. (2009). Lessons from nearly 20 years of precision agriculture research, development, and adoption as a guide to its appropriate application. *Crop and Pasture Science, 60*(3), 197–217.

Brey, P. (2018). The strategic role of technology in a good society. *Technology in Society, 52*, 39–45.

Bridle, J. (2018). *New dark age: Technology and the end of the future*. Verso.

Brown, S. L., Chaney, R. L., & Hettiarachchi, G. M. (2016). Lead in urban soils: A real or perceived concern for urban agriculture? *Journal of Environmental Quality, 45*(1), 26–36.

Brudermann, T., & Sangkakool, T. (2017). Green roofs in temperate climate cities in Europe–an analysis of key decision factors. *Urban Forestry & Urban Greening, 21*, 224–234.

Bruinsma, J. (2003). *World agriculture: Towards 2015/2030: A FAO perspective*. Earthscan/Food and Agriculture Organization.

Burgoine, T., Mackenbach, J. D., Lakerveld, J., Forouhi, N. G., Griffin, S. J., Brage, S., Wareham, N. J., & Monsivais, P. (2017). Interplay of socioeconomic status and supermarket distance is associated with excess obesity risk: A UK cross-sectional study. *International Journal of Environmental Research and Public Health, 14*(11), 1290.

Burt, K. G., Burgermaster, M., D'Alessandro, D., Paul, R., & Stopler, M. (2020). New York city fourth graders who receive a climate change curriculum with hydroponic gardening have higher science achievement scores. *Applied Environmental Education & Communication, 19*(4), 402–414.

Busby, M. (2020). How coronavirus has led to a UK boom in community food growing. *The Guardian*. Available at: https://www.theguardian.com/world/2020/aug/24/how-coronavirus-has-led-to-a-uk-boom-in-community-food-growing. Accessed 10.07.2021.

Butler. (1962). *Hydroponics as a hobby – Growing plants without soil*. University of Illinois. Available at https://www.ideals.illinois.edu/bitstream/handle/2142/33041/1129561.pdf. Accessed 02.04.2020.

Campos-Vega, R., Loarca-Pina, G., Vergara-Castaneda, H. A., & Oomah, B. D. (2015). Spent coffee grounds: A review on current research and future prospects. *Trends in Food Science & Technology, 45*(1), 24–36.

Capra, F. (1996). *The web of life: A new synthesis of mind and matter*. HarperCollins.

Caputo, S., Schwab, E., & Tsiambaos, K. (2016). Emergent approaches to urban gardening. In *Urban allotment gardens in Europe* (pp. 229–253). Routledge.

Caputo, S., Rumble, H., & Schaefer, M. (2020). "I like to get my hands stuck in the soil": A pilot study in the acceptance of soil-less methods of cultivation in community gardens. *Journal of Cleaner Production, 258*, 120585.

Caputo, S., Schoen, V., Specht, K., Grard, B., Blythe, C., Cohen, N., Fox-Kämper, R., Hawes, J., Newell, J., & Poniży, L. (2021). Applying the food-energy-water nexus approach to urban agriculture: From FEW to FEWP (food-energy-water-people). *Urban Forestry & Urban Greening, 58*, 126934.

Carlos, A., Espinal, C. A., & Matulić, D. (2019). Recirculating aquaculture technologies. In S. Goddek, A. Alyssa Joyce, B. Kotzen, & G. M. Burnell (Eds.), *Aquaponics food production systems: Combined aquaculture and hydroponic production Technologies for the Future*. Springer.

Carson, R. (2000:1962). *Silent spring*. Penguin.

Carson, S. (2002). *Silent spring*. Houghton Mifflin Harcourt.

Cattivelli, V. (2020). The motivation of urban gardens in mountain areas. The case of South Tyrol. *Sustainability, 12*, 4304. https://doi.org/10.3390/su12104304

Certomà, C., & Tornaghi, C. (2015). Political gardening. Transforming cities and political agency. *Local Environment, 20*(10), 1123–1131.

Chai, B. C., van der Voort, J. R., Grofelnik, K., Eliasdottir, H. G., Klöss, I., & Perez-Cueto, F. J. (2019). Which diet has the least environmental impact on our planet? A systematic review of vegan, vegetarian and omnivorous diets. *Sustainability, 11*(15), 4110.

Chang, S. T. (1977). The origin and early development of straw mushroom cultivation. *Economic Botany, 31*(3), 374–376.

Chang, S. T. (2009). *Training manual on mushroom cultivation technology*. United Nations-Asian And Pacific Centre For Agricultural Engineering And Machinery (UN-APCAEM).

Chang, M., & Morel, K. (2018). Reconciling economic viability and socio-ecological aspirations in London urban microfarms. *Agronomy for Sustainable Development, 38*(1), 9.

Chen, P., Zhu, G., Kim, H. J., Brown, P. B., & Huang, J. Y. (2020). Comparative life cycle assessment of aquaponics and hydroponics in the Midwestern United States. *Journal of Cleaner Production, 275*, 122888.

Cinquemani. (2019). *Italy's young farmers: Networking to make revenues*. EURACTIV. Available at https://www.euractiv.com/section/agriculture-food/news/italys-young-farmers-networking-for-making-revenues/. Accessed 05.01.2021.

Clark, K. H., & Nicholas, K. A. (2013). Introducing urban food forestry: A multifunctional approach to increase food security and provide ecosystem services. *Landscape Ecology, 28*(9), 1649–1669.

Climate Change Committee. (2020). *The Sixth Carbon Budget-Agriculture and land use, land use change and forestry*.

Cohen, A., Malone, S., Morris, Z., Weissburg, M., & Bras, B. (2018). Combined fish and lettuce cultivation: An aquaponics life cycle assessment. *Procedia Cirp, 69*, 551–556.

Colasanti, K. J. A., Hamm, M. W., & Litjens, C. M. (2012). The City as an "agricultural powerhouse"? Perspectives on expanding urban agriculture from Detroit, Michigan. *Urban Geography, 33*(3), 348–369.

Colella, C., Carradore, R., & Cerroni, A. (2019). Problem setting and problem solving in the case of olive quick decline syndrome in Apulia, Italy: A sociological approach. *Phytopathology, 109*(2), 187–199.

Collin, F., & Pedersen, D. B. (2015). The Frankfurt school, science and technology studies, and the humanities. *Social Epistemology, 29*(1), 44–72.

Commission Regulation. (2009). (EC) no 710/2009 of 5 august 2009 amending regulation (EC) no. 889/2008 laying down detailed rules for the implementation of council regulation (EC) no. 834/2007, as regards laying down detailed rules on organic aquaculture animal and seaweed production. *Official Journal of European Union Law, 204*, 15–34.

Commission Regulation (EU). 2018/848 of the European Parliament and of the Council of 30 May 2018 on organic production and labelling of organic products and repealing Council Regulation (EC) No 834/2007.

Corner, A., Parkhill, K., Pidgeon, N., & Vaughan, N. E. (2013). Messing with nature? Exploring public perceptions of geoengineering in the UK. *Global Environmental Change, 23*(5), 938–947.

Cottam, H., & Leadbeater, C. (2004). *RED paper 01: Health: Co-creating services*. Design Council.

Council Regulation (EC) No 834/2007 of 28 June 2007 on organic production and labelling of organic products and repealing Regulation (EEC) No 2092/91.

Cox, S., & Slegers, M. (2014). RotterZwam, edible mushrooms from Rotterdam. *Urban Agriculture Magazine, 28*, 52–54.

Crossley, P. L. (2004). Sub-irrigation in wetland agriculture. *Agriculture and Human Values, 21*(2/3), 191–205.

Cunha Zied, D., Sánchez, J. E., Noble, R., & Pardo-Giménez, A. (2020). Use of spent mushroom substrate in new mushroom crops to promote the transition towards a circular economy. *Agronomy, 10*(9), 1239.

Daly, H. E. (1996). *Beyond growth: The economics of sustainable development*. Beacon Press.

Datta, S. (2015). Aquaponics: Its present status and potential. *Fishing Chimes, 34*(11), 44–48.

De Amorim, W. S., Deggau, A. B., do Livramento Gonçalves, G., da Silva Neiva, S., Prasath, A. R., & de Andrade, J. B. S. O. (2019). Urban challenges and opportunities to promote sustainable food security through smart cities and the 4th industrial revolution. *Land Use Policy, 87*, 104065.

De Bon, H., Parrot, L., & Moustier, P. (2010). Sustainable urban agriculture in developing countries. A review. *Agronomy for Sustainable Development, 30*(1), 21–32.

De Schutter, O. (2012). Agroecology, a tool for the realization of the right to food. In *Agroecology and strategies for climate change* (pp. 1–16). Springer.

Despommier, D. (2010). *The vertical farm: Feeding the world in the 21st century*. Macmillan.

Dewey, C. (2017). A growing number of young Americans are leaving desk jobs to farm. *The Washington Post*, 23.

Di Lorenzo, R., Pisciotta, A., Santamaria, P., & Scariot, V. (2013). From soil to soil-less in horticulture: Quality and typicity. *Italian Journal of Agronomy, 8*(4), 30.

Di Marco, P., Petochi, T., Marino, G., Priori, A., Finoia, M. G., Tomassetti, P., Porrello, S., Giorgi, G., Lupi, P., Bonelli, A., & Parisi, G. (2017). Insights into organic farming of European sea bass Dicentrarchus labrax and gilthead sea bream Sparus aurata through the assessment of environmental impact, growth performance, fish welfare and product quality. *Aquaculture, 471*, 92–105.

Digital Europe. (2010). *A transformational agenda for the digital age – Digital Europe's vision 2020*. Available at https://digitaleurope.blogactiv.eu/2010/08/26/a-transformational-agenda-for-the-digital-age/. Accessed 12.10.2019.

Dimitri, C., Oberholtzer, L., & Pressman, A. (2016). Urban agriculture: Connecting producers with consumers. *British Food Journal, 118*(3), 603–617.

Diver, S. (2006) *Aquaponics – Integration of hydroponics with aquaculture*. ATTRA—National Sustainable Agriculture Information Service. Available at https://attra.ncat.org/attra-pub/download.php?id=56. Downloaded 07-08-2019.

Dorr, E., Koegler, M., Gabrielle, B., & Aubry, C. (2020). Life cycle assessment of a circular, urban mushroom farm. *Journal of Cleaner Production, 125668*.

Dos Santos, M. J. P. L. (2018). Nowcasting and forecasting aquaponics by Google trends in European countries. *Technological Forecasting and Social Change, 134*, 178–185.

Doshi, V.(2020, March 5). Inside agri-valley – Backlash as Dutch go from famine to feast. *The Guardian*.

Easin, M. N., Ahmed, R., Alam, M. S., Reza, M. S., & Ahmed, K. U. (2017). Mushroom cultivation as a small-scale family enterprise for the alternative income generation in rural Bangladesh. *International Journal of Agriculture, Forestry and Fisheries, 5*(1), 1–8.

EC – European Commission. (2020a). *Commission Implementing Regulation (EU) 2020/1201 of 14 August 2020*. Available at https://ec.europa.eu/food/plant/plant_health_biosecurity/legislation/emergency_measures/xylella-fastidiosa_en. Accessed 13.09.2020.

EC – European Commission. (2020b). *Horizon 2020 – Work Programme 2018–2020; 16*. Science with and for Society. European Commission.

Espinal, C. A., & Matulić, D. (2019). Recirculating aquaculture technologies. In *Aquaponics food production systems* (pp. 35–76). Springer.

Etzkowitz, H., & Leydesdorff, L. (1995). The Triple helix–University-industry-government relations: A laboratory for knowledge based economic development. *EASST Review, 14*(1), 14–19.

European Commission. (2014). *Public perceptions of science, research and innovation*. Special Eurobarometer 419.

Eurostat. (2015). *Farm structure survey 2013 – Main results*. Available at http://ec.europa.eu/eurostat/statistics-explained/index.php/Farm_structure_survey_2013_-_main_results. Accessed 26.02.2020.

Eurostat. (2019). https://ec.europa.eu/eurostat/statistics-explained/index.php/Performance_of_the_agricultural_sector#Value_of_agricultural_output. Accessed 16.09.2020.

Fall, S. T., & de Zeeuw, H. (n.d.) *Appropriate methods for technology development in urban agriculture*. Available at https://citeseerx.ist.psu.edu/viewdoc/download?doi=10.1.1.729.2370&rep=rep1&type=pdf. Accessed 10.12.2019.

FAO. (1957). *Fish culture in rice fields*. In A preliminary review and annotated bibliography from Section II and III of Proc. Indo-Pacif. Fish. Coun. 7. FAO, Rome, pp. 193–206

FAO. (2011). *The state of the world's land and water resources for food and agriculture (SOLAW) – Managing systems at risk*. Food and Agriculture Organization of the United Nations/Earthscan.

FAO. (2014). *Growing Greener cities in Latin America and the Caribbean: An FAO report on Urban and Peri-urban agriculture in the Region*. Food & Agriculture Org.

FAO. (2015). *Deep Water culture aquaponic unit: step by step description*.

FAO. (2017). *Indoor oyster mushroom cultivation for livelihood diversification and increased resilience in Uganda*. Available at http://www.fao.org/3/CA2568EN/ca2568en.pdf. Accessed 10.20.2020.

FAO. (2018). *World food and agriculture – Statistical pocketbook 2018*. FAO.

FAO. (2019). *FAO framework for the Urban Food Agenda Leveraging sub-national and local government action to ensure sustainable food systems and improved nutrition*.

FAO. (2020a). *The state of world fisheries and aquaculture 2020*. Sustainability in action. Rome.

FAO, Maruland, C., & Izquierdo, J. (1997). *Popular hydroponic gardens – Technical manual*. FAO.

Farran, I., & Mingo-Castel, A. M. (2006). Potato minituber production using aeroponics: Effect of plant density and harvesting intervals. *American Journal of Potato Research, 83*(1), 47–53. Accessed 21.6.2020.

Fecondini, M., Damasio de Faria, A. C., Michelon, N., Mezzetti, M., Orsini, F., & Gianquinto, G. (2009a). *Learning the value of gardening: results from an experience of community based simplified hydroponics in north-east Brazil*. In II International conference on landscape and urban horticulture 881 (pp. 111–116).

Fecondini, M., Michelon, N., Gianquinto, G., Orsini, F., & Mezzetti, M. (2009b). *Simplified substrate soilless culture for vegetable production in Trujillo, Peru*. In II International conference on landscape and urban horticulture 881 (pp. 163–167).

Fecondini, M., Damasio de Faria, A. C., Michelon, N., Mezzetti, M., Orsini, F., & Gianquinto, G. (2010). Learning the value of gardening: Results from an experience of community based simplified hydroponics in north-East Brazil. *Acta Horticulturae, 881*, 111–116.

Feenberg, A. (2017). A critical theory of technology. In U. Felt, R. Fouché, C. A. Miller, & L. Laurel Smith-Doerr (Eds.), *Handbook of science and technology studies* (pp. 635–663). MIT Press.

Felt, U., Wynne, B., Callon, M., Gonçalves, M.E., Jasanoff, S., Jepsen, M., Joly, P.B., Konopasek, Z., May, S., Rip, A., & Neubauer, C., (2007) *Taking European knowledge society seriously*. Report of the expert group on science and governance to the Science, Economy and Society Directorate, Directorate-General for Research, European Commission. Office des Publications Officielles des Communautés Européennes.

Fischer-Kowalski, M. (1998). Society's metabolism: The intellectual history of materials flow analysis, part I, 1860–1970. *Journal of Industrial Ecology, 2*(1), 61–78.

Fitzgerald, D. (1991). Beyond tractors: The history of technology in American agriculture. *Technology and Culture, 32*(1), 114–126.

Flyvbjerg, B. (2006). Five misunderstandings about case-study research. *Qualitative Inquiry, 12*(2), 219–245.

Fontes, M. R. (1973). Controlled-environment horticulture in the Arabian Desert at Abu Dhabi. *Horticultural Science, 8*, 13–16.

Food Ethics Council. (2015). *From hydroponics to underground farms*. Available at https://www.foodethicscouncil.org/app/uploads/160705%20From%20hydroponics%20to%20underground%20farms_FINAL.pdf. Accessed on 10.10.2019.

Foteinis, & Chatzisymeon, E. (2016). Life cycle assessment of organic versus conventional agriculture. A case study of lettuce cultivation in Greece. *Journal of Cleaner Production, 112*, 2462–2471.

Francis, C., Lieblein, G., Gliessman, S., Breland, T. A., Creamer, N., Harwood, R., Salomonsson, L., Helenius, J., Rickerl, D., Salvador, R., & Wiedenhoeft, M. (2003). Agroecology: The ecology of food systems. *Journal of Sustainable Agriculture, 22*(3), 99–118.

Fraser, L. K., & Edwards, K. L. (2010). The association between the geography of fast food outlets and childhood obesity rates in Leeds, UK. *Health & Place, 16*(6), 1124–1128.

Fremstad, A., & Paul, M. (2020). Opening the farm gate to women? The gender gap in U.S. agriculture. *Journal of Economic Issues, 54*(1), 124–141. https://doi.org/10.1080/00213624.2020.1720569

Fuglie, K. O., & Wang, S. L. (2012). Productivity growth in global agriculture shifting to developing countries. *Choices, 27*(316–2016-7633).

Gallup. (2019). *Wellcome global monitor – First wave findings*. Wellcome.

Galluzzi, G., Eyzaguirre, P., & Negri, V. (2010). Home gardens: Neglected hotspots of agrobiodiversity and cultural diversity. *Biodiversity and Conservation, 19*(13), 3635–3654.

García-Sempere, A., Hidalgo, M., Morales, H., Ferguson, B. G., Nazar-Beutelspacher, A., & Rosset, P. (2018). Urban transition toward food sovereignty. *Globalizations, 15*(3), 390–406.

Garg, A., & Balodi, R. (2014). Recent trends in agriculture: Vertical farming and organic farming. *Adv Plants Agric Res, 1*(4), 00023.

Geels, F. W. (2004). From sectoral systems of innovation to socio-technical systems: Insights about dynamics and change from sociology and institutional theory. *Research Policy, 33*(6–7), 897–920.

Gericke, W. F. (1940) The complete guide to soilless gardening. In *The complete guide to soilless gardening*. Prentice-Hall.

Germer, J., Sauerborn, J., Asch, F., de Boer, J., Schreiber, J., Weber, G., & Müller, J. (2011). Skyfarming an ecological innovation to enhance global food security. *Journal für Verbraucherschutz und Lebensmittelsicherheit, 6*(2), 237–251.

Ghose, R., & Pettygrove, M. (2014). Urban community gardens as spaces of citizenship. *Antipode, 46*(4), 1092–1112.

Ghosh, S. (2014). Measuring sustainability performance of local food production in home gardens. *Local Environment, 19*(1), 33–55.

Giorda, E. (2012). Farming in Mowtown: Competing narratives for urban development and urban agriculture in Detroit. In *Sustainable food planning: Evolving theory and practice*. Wageningen Academic Publishers.

Girardet, H. (2004). *Cities, people, planet. Liveable cities for a sustainable world*. Wiley.

Gittleman, M., Jordan, K., & Brelsford, E. (2012). Using citizen science to quantify community garden crop yields. *Cities and the Environment (CATE), 5*(1), 4.

Giusti, M., Barthel, S., & Marcus, L. (2014). Nature routines and affinity with the biosphere: A case study of preschool children in Stockholm. *Children, Youth and Environments*.

Glavan, M., Schmutz, U., Williams, S., Corsi, S., Monaco, F., Kneafsey, M., Rodriguez, P. A. G., Čenič-Istenič, M., & Pintar, M. (2018). The economic performance of urban gardening in three European cities–examples from Ljubljana, Milan and London. *Urban Forestry & Urban Greening, 36*, 100–122.

Goddek, S., Delaide, B., Mankasingh, U., Ragnarsdottir, K. V., Jijakli, H., & Thorarinsdottir, R. (2015). Challenges of sustainable and commercial aquaponics. *Sustainability, 7*(4), 4199–4224.

Goddek, S., Espinal, C., Delaide, B., Jijakli, M., Schmautz, Z., Wuertz, S., & Keesman, K. (2016). Navigating towards decoupled aquaponic systems: A system dynamics design approach. *Water, 8*, 303.

Goddek, S., Joyce, A., Kotzen, B., & Burnell, G. M. (Eds.). (2019a). *Aquaponics food production systems: Combined aquaculture and hydroponic production Technologies for the Future* (p. 619). Springer.

Goddek, S., Joyce, A., Wuertz, S., Körner, O., Bläser, I., Reuter, M., & Keesman, K. J. (2019b). Decoupled aquaponics systems. In *Aquaponics food production systems* (pp. 201–229). Springer.

Goldstein, B., Hauschild, M., Fernandez, J., & Birkved, M. (2016). Testing the environmental performance of urban agriculture as a food supply in northern climates. *Journal of Cleaner Production, 135*, 984–994.

Gómez-Baggethun, E., De Groot, R., Lomas, P. L., & Montes, C. (2010). The history of ecosystem services in economic theory and practice: From early notions to markets and payment schemes. *Ecological Economics, 69*(6), 1209–1218.

González, A., Cruz, M., Losoya, C., Nobre, C., Loredo, A., Rodríguez, R., Contreras, J., & Belmares, R. (2020). Edible mushrooms as a novel protein source for functional foods. *Food & Function, 11*(9), 7400–7414.

Gowdy, J., & Baveye, P. (2019). An evolutionary perspective on industrial and sustainable agriculture. In *Agroecosystem diversity* (pp. 425–433). Academic.

Grand View Research. (2020a). *Hydroponics market size, share & trends analysis report by type (aggregate systems, liquid systems), By crops (tomatoes, lettuce, peppers, cucumbers, herbs), by region, and segment forecasts, 2019–2025*. Available at https://www.grandviewresearch.com/industry-analysis/hydroponics-market. Accessed 21.09.2020.

Grard, B., Bel, N., Marchal, N., Madre, N., Castell, J. F., Cambier, P., Houot, S., Manouchehri, N., Besancon, S., Michel, J. C., & Chenu, C. (2015). Recycling urban waste as possible use for rooftop vegetable garden. *Future of Food: Journal on Food, Agriculture and Society, 3*(1), 21–34.

Grard, B. J., Manouchehri, N., Aubry, C., Frascaria-Lacoste, N., & Chenu, C. (2020). Potential of technosols created with urban by-products for rooftop edible production. *International Journal of Environmental Research and Public Health, 17*(9), 3210.

Graves, C. G. (1983). The Nutrient Film Technique. *Horticultural Review, 5*, 1–44.

Grebitus, C., Printezis, I., & Printezis, A. (2017). Relationship between consumer behavior and success of urban agriculture. *Ecological Economics, 136*, 189–200.

Greensgrow. (2020). *What is urban farming?*https://www.greensgrow.org/urban-farm/what-is-urban-farming/. Accessed 06.03.2020.

Gren, Å., & Andersson, E. (2018). Being efficient and green by rethinking the urban-rural divide—Combining urban expansion and food production by integrating an ecosystem service perspective into urban planning. *Sustainable Cities and Society*. https://doi.org/10.1016/j.scs.2018.02.031

Grimm, D., & Wösten, H. A. (2018). Mushroom cultivation in the circular economy. *Applied Microbiology and Biotechnology, 102*(18), 7795–7803.

Gruda, N., & Tanny, J. (2014). Chapter 10: Protected crops. In: G. R. Dixon & D. E. Aldous (Eds.), *Horticulture – plants for people and places,Production Horticulture* (Vol 1, pp 327–405). Springer.

Gunady, M. G. A., Biswas, W., Solah, V. A., & James, A. P. (2012). Evaluating the global warming potential of the fresh produce supply chain for strawberries, romaine/cos lettuces (Lactuca sativa), and button mushrooms (Agaricus bisporus) in Western Australia using life cycle assessment (LCA). *Journal of Cleaner Production, 28*, 81e87.

Guo, X., Broeze, J., Groot, J. J., Axmann, H., & Vollebregt, M. (2020). A worldwide hotspot analysis on food loss and waste, associated greenhouse gas emissions, and protein losses. *Sustainability, 12*(18), 7488.

Halais, F. (2014). Can We Do Full-Scale Farming Inside City Centers? *Citylab*. Available at https://www.citylab.com/life/2014/08/can-urban-agriculture-work-on-a-commercial-scale/378984/. Accessed 27.03.2020.

Halwart, M., & Gupta, M. V. (Eds.). (2004). *Culture of fish in rice fields*. FAO.

Hanko, J. (2001). *Mushroom cultivation for people with disabilities–a training manual*. Food and Agriculture Organization of the United Nations, Regional Office for Asia and the Pacific.

Hara, Y., McPhearson, T., Sampei, Y., & McGrath, B. (2018). Assessing urban agriculture potential: A comparative study of Osaka, Japan and New York city. *United States. Sustainability Science, 13*(4), 937–952.

Hardman, M., Adams, M., Armitage, R., Cryer, S., & Stein, S. (n.d.). *High tech sustainable food growing options study*. University of Salford.

Hart, H. (1930). Nicolas Theodore De Saussure. *Plant Physiology, 5*(3), 424–429. Available at https://www.ncbi.nlm.nih.gov/pmc/articles/PMC440232/

Hedenus, F., Wirsenius, S., & Johansson, D. J. (2014). The importance of reduced meat and dairy consumption for meeting stringent climate change targets. *Climatic Change, 124*(1–2), 79–91.

Henley, J. (2020). The future of food: inside the world's largest urban farm – built on a rooftop. *The Guardian*, 8 July 2020.

Hershey, D. R. (1994). Solution culture hydroponics: History & inexpensive equipment. *The American Biology Teacher, 56*(2), 111–118.

Higgins, C., Margot, H., Warnquist, S., Obeysekare, E., & Mehta, K. (2017, October). Mushroom cultivation in the developing world: a comparison of cultivation technologies. In *2017 IEEE Global Humanitarian Technology Conference (GHTC)* (pp. 1–7). IEEE.

Higgs, E. S. (1976). The history of European agriculture – The uplands. *Philosophical Transactions of the Royal Society of London. B, Biological Sciences, 275*(936), 159–173.

Hindle, R. L. (2012). A vertical garden: Origins of the vegetation-bearing architectonic structure and system (1938). *Studies in the History of Gardens & Designed Landscapes, 32*(2), 99–110.

Hirsch, D., Meyer, C., Klement, J., Hamer, M., & Terlau, W. (2016). Urban agriculture and food systems dynamics: Urban gardening and urban farming of the Bonn-Rhein-Sieg region, Germany. *Proceedings in Food System Dynamics*, 406–419.

Hoagland, D. R., & Arnon, D. I. (1950). *The water-culture method for growing plants without soil*. College of Agriculture/University of California.

Hoevenaars, K., Junge, R., Bardocz, T., & Leskovec, M. (2018). EU policies: New opportunities for aquaponics. *Ecocycles, 4*(1), 10–15.

Holland, L. (2004). Diversity and connections in community gardens: A contribution to local sustainability. *Local Environment, 9*(3).

Hoover, B. (2013). White spaces in black and Latino places: Urban agriculture and food sovereignty. *Journal of Agriculture, Food Systems, and Community Development, 3*(4), 109–115.

Hopkins, R. (2008). *The transition handbook*. Green Books.

Horrigan, L., Lawrence, R. S., & Walker, P. (2002). How sustainable agriculture can address the environmental and human health harms of industrial agriculture. *Environmental Health Perspectives, 110*(5), 445–456.

Horst, M., McClintock, N., & Hoey, L. (2017). The intersection of planning, urban agriculture, and food justice: A review of the literature. *Journal of the American Planning Association, 83*(3), 277–295.

Hortidaily. (2019, October 14). *Inauguration of ALTMARKTgarten Oberhausen experiments regarding urban agriculture*. Available at https://www.hortidaily.com/article/9148894/experiments-regarding-urban-agriculture/. Accessed on 12.11.2020.

Hossain, M. (2017). Mapping the frugal innovation phenomenon. *Technology in Society, 51*, 199–208.

Hungwe, C. (2006). Urban agriculture as a survival strategy. An analysis of the activities of Bulawayo and Gweru urban farmers, Zimbabwe. *Urban Agriculture Notes*, 2.

International Center for Living Aquatic Resources Management, & International Institute of Rural Reconstruction. (2001). *Integrated agriculture-aquaculture: a primer* (No. 407). Food & Agriculture Org.. Available at http://www.fao.org/3/Y1187E/Y1187E00.htm. Accessed 06-08-2019.

Ioannou, B., Moran, N., Sondermann, M., Certomà, C., Hardman, M., Anthopoulou, T., Notteboom, B., & Silvestri, G. (2016). Grassroots gardening movements: Towards cooperative forms of green urban development? In S. Bell, R. Fox-Kamper, M. Benson, S. Caputo, N. Kaeshavarz, S. Noori, & A. Voigt (Eds.), *Urban allotment gardens in Europe* (pp. 62–90). Routledge.

Ipsos Mori. (2014). *Public attitudes to science 2014*. Available at https://www.ipsos.com/ipsos-mori/en-uk/public-attitudes-science-2014. Accessed 11.12.2021.

Irfanullah, H. M., Azad, M. A. K., Kamruzzaman, M., & Wahed, M. A. (2011). *Floating gardening in Bangladesh: a means to rebuild lives after devastating flood*. Available at http://nopr.niscair.res.in/handle/123456789/11064. Accessed 06-08-2019.

Irwin, A., & Wynne, B. (Eds.). (1996). *Misunderstanding science? The public reconstruction of science and technology*. Cambridge University Press.

Islam, T., & Atkins, P. (2007). Indigenous floating cultivation: A sustainable agricultural practice in the wetlands of Bangladesh. *Development in Practice, 17*(1), 130–136.

Izquierdo, J. (2005). *Simplified hydroponics: a tool for food security in Latin America and the Caribbean*. In International conference and exhibition on soilless culture: ICESC 2005 *742* (pp. 67–74).

Jensen, M. H. (1997a). Hydroponics. *Horticultural Science, 32*(6), 1018–1021.

Jensen, M. H. (1997b). Hydroponics worldwide. In *International symposium on growing media and hydroponics 481* (pp. 719–730).

Jensen, M. H., & Teran, M. A. (1971). Use if controlled environment for vegetable production in desert regions of the world. *Horticultural Science, 6*, 33–36.

Jim, C. Y. (2017). An archaeological and historical exploration of the origins of green roofs. *Urban Forestry & Urban Greening, 27*, 32–42.

Jones, S. (2002). Evolution of aquaponics. *Aquaponics J, 6*(1).

Jones, J. B., Jr. (2014). *Complete guide for growing plants hydroponically*. CRC Press.

Junge, R., König, B., Villarroel, M., Komives, T., & Jijakli, M. H. (2017). Strategic points in aquaponics. *Water, 9*, 182.

Junge, R., Bulc, T. G., Anseeuw, D., Yildiz, H. Y., & Milliken, S. (2019). Aquaponics as an educational tool. *Aquaponics Food Production Systems, 561*.

Kadhila-Muandingi, N. P., Mubiana, F. S., & Halueendo, K. L. (2008). *Mushroom cultivation: A beginners guide*. University of Namibia.

Kalantari, F., Tahir, O. M., Joni, R. A., & Fatemi, E. (2018). Opportunities and challenges in sustainability of vertical farming: A review. *Journal of Landscape Ecology, 11*(1), 35–60.

Kangmin, L. (1988). Rice-fish culture in China: A review. *Aquaculture, 71*(3), 173–186.

Karanja, N., Yeudall, F., Mbugua, S., Njenga, M., Prain, G., Cole, D. C., Webb, A. L., Sellen, D., Gore, C., & Levy, J. M. (2010). Strengthening capacity for sustainable livelihoods and food security through urban agriculture among HIV and AIDS affected households in Nakuru, Kenya. *International Journal of Agricultural Sustainability, 8*(1–2SI), 40–53.

Kashangura, C., Kunjeku, E. C., Chirara, T., Mabveni, A. R. S., Mswaka, A., & Dalu, V. (2005). *Manual for mushroom cultivation (especially for growers with limited financial and material resources)*. Biotechnology Trust of Zimbabwe.

Khush, G. S. (2001). Green revolution: The way forward. *Nature Reviews Genetics, 2*(10), 815–822.

Kim, S. (2018). Managing millennials' personal use of technology at work. *Business Horizons, 61*(2), 261–270.

Kirby, C. K., Specht, K., Fox-Kämper, R., Hawes, J. K., Cohen, N., Caputo, S., Ilieva, R.T., Lelievre, A., Poniży, L., Schoen, V. & Blythe, C. (2021) Differences in motivations and social impacts across urban agriculture types: Case studies in Europe and the US. *Landscape and Urban Planning, 212*, 104110.

Kitao, Y. (2005). *Collective urban design: Shaping the city as a collaborative process*. Delft University Press.

Kitaya, Y., Hirai, H., Wei, X., Islam, A. F. M. S., & Yamamoto, M. (2008). Growth of sweet potato cultured in the newly designed hydroponic system for space farming. *Advances in Space Research, 41*(5), 730–735.

Kledal, P. R., König, B., & Matulić, D. (2019). Aquaponics: The ugly duckling in organic regulation. In *Aquaponics food production systems* (Vol. 487). Springer.

Knaus, U., & Palm, H. W. (2017). Effects of fish biology on ebb and flow aquaponical cultured herbs in northern Germany (Mecklenburg Western Pomerania). *Aquaculture, 466*, 51–63.

Kotzen, B., Emerenciano, M. G. C., Moheimani, N., & Burnell, G. M. (2019). Aquaponics: Alternative types and approaches. In *Aquaponics food production systems* (pp. 301–330). Springer.

Koutrotsios, G., Kalogeropoulos, N., Kaliora, A. C., & Zervakis, G. I. (2018). Toward an increased functionality in oyster (Pleurotus) mushrooms produced on grape marc or olive mill wastes serving as sources of bioactive compounds. *Journal of Agricultural and Food Chemistry, 66*(24), 5971–5983.

Kratky, B. A., Maehira, G. T., Cupples, R. J., & Bernabe, C. C. (2005). *Non-circulating hydroponic methods for growing tomatoes*. In Proceedings of the national agricultural plastics congress (Vol. 32, pp. 31–36).

Kriewald, S., Pradhan, P., Costa, L., Ros, A. G. C., & Kropp, J. P. (2019). Hungry cities: How local food self-sufficiency relates to climate change, diets, and urbanisation. *Environmental Research Letters, 14*(9), 094007.

Krikser, T., Piorr, A., Berges, R., & Opitz, I. (2016). Urban agriculture oriented towards self-supply, social and commercial purpose: A typology. *Land, 5*(3), 28.

Krusenvik, L. (2016). *Using case studies as a scientific method: Advantages and disadvantages*. Available at https://www.diva-portal.org/smash/get/diva2:1054643/FULLTEXT01.pdf. Accessed 20.11.2020.

Lang, M. (2020a). Consumer acceptance of blending plant-based ingredients into traditional meat-based foods: Evidence from the meat-mushroom blend. *Food Quality and Preference, 79*, 103758.

Lang, T. (2020b). *Feeding Britain: Our food problems and how to fix them*. Penguin.

Langemeyer, J., Latkowska, M. J., Gómez-Baggethun, E. N., Voigt, A., Calvet-Mir, L., Pourias, J., Camps-Calvet, M., Orsini, F., Breuste, J., Artmann, M., & Jokinen, A. (2016). Ecosystem services from urban gardens. *Urban allotment gardens in Europe*, 115–141.

Lauri, A. (2017) *The 'companization' organizational model. Managerial challenges and responses in a peculiar hybrid organization. A case study on Plantagon International*. Master THESIS – LUISS, Milano.

Leckie, G. J. (1996). 'They never trusted me to Drive': Farm girls and the gender relations of agricultural information transfer. *Gender, Place and Culture: A Journal of Feminist Geography, 3*(3), 309–326. https://doi.org/10.1080/09663699625586

Leiva, F. J., Saenz-Díez, J. C., Martínez, E., Jimenez, E., & Blanco, J. (2015). Environmental impact of Agaricus bisporus cultivation process. *European Journal of Agronomy, 71*, 141e148.

Lemaire, A., & Limbourg, S. (2019). How can food loss and waste management achieve sustainable development goals? *Journal of Cleaner Production, 234*, 1221–1234.

Lennard W. A. (2005). *Aquaponic integration of Murray Cod (Maccullochella peelii peelii) aquaculture and lettuce (Lactuca sativa) hydroponics*. Thesis (Ph.D.). RMIT University.

Lennard, W. (2012). *Aquaponics system design parameters: Basic system water chemistry*. Aquaponic Solutions.

Lennard, W., & Goddek, S. (2019). Aquaponics: The basics. In S. Goddek, A. Alyssa Joyce, B. Kotzen, & G. M. Burnell (Eds.), *Aquaponics food production systems: Combined aquaculture and hydroponic production technologies for the future*. Springer.

Lennard, W. A., & Leonard, B. V. (2004). A comparison of reciprocating flow versus constant flow in an integrated, gravel bed, aquaponic test system. *Aquaculture International, 12*(6), 539–553.

Lewis, W. M., Yopp, J. H., Schramm, H. L., Jr., & Brandenburg, A. M. (1978). Use of hydroponics to maintain quality of recirculated water in a fish culture system. *Transactions of the American Fisheries Society, 107*(1), 92–99.

Little, N. G., McCoy, T., Wang, C., & Dill, S. P. (2019). Results of a needs assessment of urban farmers in Maryland. *Journal of the NACAA, 12*(1).

Lobillo-Eguíbar, J., Fernández-Cabanás, V. M., Bermejo, L. A., & Pérez-Urrestarazu, L. (2020). Economic sustainability of small-scale aquaponic systems for food self-production. *Agronomy, 10*(10), 1468.

Lohrberg, F., Lička, L., Scazzosi, L., & Timpe, A. (Eds.). (2016). *Urban agriculture Europe*. Jovis.

Love, D. C., Uhl, M. S., & Genello, L. (2015a). Energy and water use of a small-scale raft aquaponics system in Baltimore, Maryland, United States. *Aquaculture Engineering, 68*, 19–27.

Love, D. C., Fry, J. P., Li, X., Hill, E. S., Genello, L., Semmens, K., & Thompson, R. E. (2015b). Commercial aquaponics production and profitability: Findings from an international survey. *Aquaculture, 435*, 67–74.

Lowder, S. K., Skoet, J., & Raney, T. (2016). The number, size, and distribution of farms, smallholder farms, and family farms worldwide. *World Development, 87*, 16–29.

MacKenzie, D., & Wajcman, J. (1999). Introductory essay: The social shaping of technology. In D. MacKenzie & J. Wajcman (Eds.), *The social shaping of technology*. Open University Press.

Mackowiak, C. L., Wheeler, R. M., Stutte, R. M., Yorio, R. M., & Ruffe, L. M. (1998). A recirculating hydroponic system for studying peanut (Arachis hypogaea L.). *HortScience, 33*, 650–651.

Malberg Dyg, P., Christensen, S., & Peterson, C. J. (2020). Community gardens and wellbeing amongst vulnerable populations: A thematic review. *Health Promotion International, 35*(4), 790–803.

Manzini, E. (2014). Making things happen: Social innovation and design. *Design Issues, 30*(1), 57–66.

Marulanda, C., & Izquierdo, J. (1993). *Popular hydroponic gardens*. Audio visual course.

May, D., Arancibia, S., Behrendt, K., & Adams, J. (2019). Preventing young farmers from leaving the farm: Investigating the effectiveness of the young farmer payment using a behavioural approach. *Land Use Policy, 82*, 317–327.

Maye, D. (2019). 'Smart food city': Conceptual relations between smart city planning, urban food systems and innovation theory. *City, Culture and Society, 16*, 18–24.

Mayer, G., & Paul, R. (2021). A systematic, mixed studies review of the outcomes of community garden participation related to food justice. *Local Environment, 26*(1), 17–42.

Mayer, B. K., Baker, L. A., Boyer, T. H., Drechsel, P., Gifford, M., Hanjra, M. A., Parameswaran, P., Stoltzfus, J., Westerhoff, P., & Rittmann, B. E. (2016). Total value of phosphorus recovery. *Environmental Science & Technology, 50*(13), 6606–6620.

McClements, D. J. (2019). *Future foods: How modern science is transforming the way we eat*. Springer.

McClintock, N. (2010). Why farm the city? Theorizing urban agriculture through a lens of metabolic rift. *Cambridge Journal of Regions, Economy and Society, 3*(2), 191–207.

McClintock, N. (2012). Assessing soil lead contamination at multiple scales in Oakland, California: Implications for urban agriculture and environmental justice. *Applied Geography, 35*(1–2), 460–473.

McClintock, N. (2014). Radical, reformist, and garden-variety neoliberal: Coming to terms with urban agriculture's contradictions. *Local Environment, 19*(2), 147–171.

McDougall, R., Kristiansen, P., & Rader, R. (2019). Small-scale urban agriculture results in high yields but requires judicious management of inputs to achieve sustainability. *Proceedings of the National Academy of Sciences of the United States of America, 116*(1), 129–134.

McGuinness, D. (2015). *Living the good life in Berlin's allotments – But for how long?* Available at https://www.bbc.co.uk/news/world-europe-35008648. Accessed 06.03.2020.

McWilliams, J. E. (2014). Turning the GM battleship: The tide of popular Opinion and the future of genetically modified foods. In D. Songstad, J. Hatfield, & D. Tomes (Eds.), *Convergence of food security, energy security and sustainable agriculture* (Biotechnology in agriculture and forestry) (Vol. 67). Springer. https://doi.org/10.1007/978-3-642-55262-5_6

Meenar, M., & Hoover, B. (2012). *Community food security via urban agriculture: Understanding people, place, economy, and accessibility from a food justice perspective.*

Merckx, T., & Pereira, H. M. (2015). Reshaping Agri-environmental subsidies: From marginal farming to large-scale rewilding. *Basic and Applied Ecology, 16*(2), 95–103.

Michelon, N., da Silva, D. F., de Faria, F. D., Gianquinto, G., & Orsini, F. (2006, *February). Improving yield of vegetables by using soilless micro-garden technologies in peri-urban area of North-East Brazil.* In VIII International symposium on protected cultivation in mild winter climates: advances in soil and soilless cultivation under 747 (pp. 57–65).

Miličić, V., Thorarinsdottir, R., Santos, M. D., & Hančič, M. T. (2017). Commercial aquaponics approaching the European market: To consumers' perceptions of aquaponics products in Europe. *Water, 9*(2), 80.

Millennium Ecosystem Assessment. (2003). *Ecosystems and human Well-being. A framework for assessment.* Island Press.

Monbiot, G. (2014). *Feral: Rewilding the land, the sea, and human life.* University of Chicago Press.

Monlau, F., Sambusiti, C., Ficara, E., Aboulkas, A., Barakat, A., & Carrère, H. (2015). New opportunities for agricultural digestate valorization: Current situation and perspectives. *Energy & Environmental Science, 8*(9), 2600–2621.

Montanari, M. (1996). *The culture of food.* Blackwell.

Moragues-Faus, A., & Morgan, K. (2015). Reframing the foodscape: The emergent world of urban food policy. *Environment and Planning A: Economy and Space, 47*(7), 1558–1573.

Morehart, C. T. (2016). Chinampa agriculture, surplus production, and political change at Xaltocan, Mexico. *Ancient Mesoamerica, 27*(1), 183–196.

Morel, K., & Léger, F. (2016). A conceptual framework for alternative farmers' strategic choices: The case of French organic market gardening microfarms. *Agroecology and Sustainable Food Systems, 40*(5), 466–492.

Morel, K., San Cristobal, M., & Léger, F. G. (2017). Small can be beautiful for organic market gardens: An exploration of the economic viability of French microfarms using MERLIN. *Agricultural Systems, 158*, 39–49.

Moreno-Peñaranda, R. (2011). *Japan's urban agriculture: Cultivating sustainability and Well-being.* United Nations.

Morgan, K. (2015). Nourishing the city: The rise of the urban food question in the global north. *Urban Studies.* https://doi.org/10.1177/0042098014534902

Mueller, N. D., Gerber, J. S., Johnston, M., Ray, D. K., Ramankutty, N., & Foley, J. A. (2012). Closing yield gaps through nutrient and water management. *Nature, 490*(7419), 254.

Murray, F., Bostock, J., & Fletcher, D. (2014). *Review of recirculation aquaculture system technologies and their commercial application.*

Murthy, P. S., & Naidu, M. M. (2012). Sustainable management of coffee industry by-products and value addition – A review. *Resources, Conservation and Recycling, 66*, 45–58.

Nadal, A., Alamús, R., Pipia, L., Ruiz, A., Corbera, J., Cuerva, E., Rieradevall, J., & Josa, A. (2017). Urban planning and agriculture. Methodology for assessing rooftop greenhouse potential of non-residential areas using airborne sensors. *Science of the Total Environment, 601*, 493–507.

Naegel, L. C. A. (1977). Combined production of fish and plants in recirculating water. *Aquaculture, 10*(1), 17–24.

Naspetti, S., & Zanoli, R. (2009). Organic food quality and safety perception throughout Europe. *Journal of Food Products Marketing, 15*(3), 249–266.

Navarro, L. M., & Pereira, H. M. (2015). Towards a European policy for rewilding. *Rewilding European Landscapes*, 205–223.

Nerantzis, E., Koliopoulos, T., & Sharma, S. (2018). Urban vertical hydroponics. *Emerging Environmental Technologies and Health Protection, 1*, 13–18.

Ngumby, E. (2020) OPINION: Scaling up urban farming beyond COVID-19. *Thomas Reuters Foundation News*. Available at https://news.trust.org/item/20200609114853-9sia1. Accessed 12.08.2021.

Niala, J. C. (2020). Dig for vitality: UK urban allotments as a health-promoting response to COVID-19. *Cities & Health*, 1–5.

Noone, S., & Jenkins, N. (2018). Digging for dementia: Exploring the experience of community gardening from the perspectives of people with dementia. *Aging & Mental Health, 22*(7), 881–888.

Oberholtzer, L., Dimitri, C., & Pressman, A. (2016). *Urban agriculture in the United States: Baseline findings of a Nationwide survey*. National Sustainable Agriculture Information Service (ATTRA) of the National Center for Appropriate Technology (NCAT).

O'Sullivan, C. A., Bonnett, G. D., McIntyre, C. L., Hochman, Z., & Wasson, A. P. (2019). Strategies to improve the productivity, product diversity and profitability of urban agriculture. *Agricultural Systems, 174*, 133–144.

Okemwa, E. (2015). Effectiveness of aquaponic and hydroponic gardening to traditional gardening. *International Journal of Scientific Research and Innovative Technology, 2*(12), 21–52.

Onofre, S. A. (2005). The floating gardens in México Xochimilco, world heritage risk site. *City & Time, 1*(3), 5.

Opitz, I., Berges, R., Piorr, A., & Krikser, T. (2016). Contributing to food security in urban areas: Differences between urban agriculture and peri-urban agriculture in the global north. *Agriculture and Human Values, 33*(2), 341–358.

Orsini, F., Kahane, R., Nono-Womdim, R., & Gianquinto, G. (2013). Urban agriculture in the developing world: A review. *Agronomy for Sustainable Development, 33*(4), 695–720.

Orsini, F., Gasperi, D., Marchetti, L., Piovene, C., Draghetti, S., Ramazzotti, S., Bazzocchi, G., & Gianquinto, G. (2014). Exploring the production capacity of rooftop gardens (RTGs) in urban agriculture: The potential impact on food and nutrition security, biodiversity and other ecosystem services in the city of Bologna. *Food Security, 6*(6), 781–792.

Owen, R., Macnaghten, P., & Stilgoe, J. (2012). Responsible research and innovation: From science in society to science for society, with society. *Science and Public Policy, 39*(6), 751–760.

Palm, H. W., Seidemann, R., Wehofsky, S., & Knaus, U. (2014). Significant factors influencing the economic sustainability of closed aquaponic systems. Part I: System design, chemo-physical parameters and general aspects. *AACL Bioflux, 7*(1), 20–32.

Palm, H. W., Knaus, U., Appelbaum, S., Goddek, S., Strauch, S. M., Vermeulen, T., Jijakli, M. H., & Kotzen, B. (2018). Towards commercial aquaponics: A review of systems, designs, scales and nomenclature. *Aquaculture International, 26*(3), 813–842.

Palm, H. W., Knaus, U., Appelbaum, S., Strauch, S. M., & Kotzen, B. (2019). Coupled aquaponics systems. In *Aquaponics food production systems* (pp. 163–199). Springer.

Paris, H. S., Janick, J., & Pitrat, M. (2008). *What the Roman emperor Tiberius grew in his greenhouses*.

Parodi, A., Leip, A., De Boer, I. J. M., Slegers, P. M., Ziegler, F., Temme, E. H., Herrero, M., Tuomisto, H., Valin, H., Van Middelaar, C. E., & Van Loon, J. J. A. (2018). The potential of future foods for sustainable and healthy diets. *Nature Sustainability, 1*(12), 782–789.

Pattnaik, B. K., & Dhal, D. (2015). Mobilizing from appropriate technologies to sustainable technologies based on grassroots innovations. *Technology in Society, 40*, 93–110.

Peluso, N. L. (2012). What's nature got to do with it? A situated historical perspective on socio-natural commodities. *Development and Change, 43*(1), 79–104.

Pereira, H. M., & Navarro, L. M. (2015). *Rewilding european landscapes* (p. 227). Springer.

Perry, J. (2017). *The history of allotments*. Available at https://www.learningwithexperts.com/gardening/blog/the-history-of-allotments. Accessed 06.03.2020.

Peters, R., & Gregory, M. (2014). Networking, small farms and EU rural development policy. *EuroChoices, 13*(1), 36–39.

Peterson, G., Allen, C. R., & Holling, C. S. (1998). Ecological resilience, biodiversity, and scale. *Ecosystems, 1*(1), 6–18.

Petts, J., & World Health Organization. (2001). *Urban agriculture in London by James Petts* (No. EUR/01/5026025 (1)). Copenhagen: WHO Regional Office for Europe.

Peyraud, J. L., & MacLeod, M. (2020). Future of EU Livestock – How to contribute to a sustainable agricultural sector. Final report. Directorate-General for Agriculture and Rural Development (European Commission): Brussels, Belgium, 82.

Pfister, S., Bayer, P., Koehler, A., & Hellweg, S. (2011). Projected water consumption in future global agriculture: Scenarios and related impacts. *Science of the Total Environment, 409*(20), 4206–4216.

Phibunwatthanawong, T., & Riddech, N. (2019). Liquid organic fertilizer production for growing vegetables under hydroponic condition. *International Journal of Recycling of Organic Waste in Agriculture, 8*(4), 369–380.

Pidgeon, N., Corner, A., Parkhill, K., Spence, A., Butler, C., & Poortinga, W. (2012). Exploring early public responses to geoengineering. *Philosophical Transactions of the Royal Society A: Mathematical, Physical and Engineering Sciences, 370*(1974), 4176–4196.

Pimbert, M. (2017). Towards a transformative urban agroecology. *Urban Agriculture Magazine, 33*, 15–17.

Pinstrup-Andersen, P. (2018). Is it time to take vertical indoor farming seriously? *Global Food Security, 17*, 233–235.

Piorr, A., Zasada, I., Doernberg, A., Zoll, F., & Ramme, W. (2018). *Research for AGRI committee-urban and Peri-urban agriculture in the EU*. European Parliament.

Poortinga, W., & Pidgeon, N. (2007). Public perceptions of agricultural biotechnology in the UK: The case of genetically modified food. *The Public, the Media and Agricultural Biotechnology*, 21–56.

Poulet, L., Fontaine, J. P., & Dussap, C. G. (2016). Plant's response to space environment: A comprehensive review including mechanistic modelling for future space gardeners. *Botany Letters, 163*(3), 337–347.

Pourias, J., Duchemin, E., & Aubry, C. (2015). Products from urban collective gardens: Food for thought or for consumption? Insights from Paris and Montreal. *Journal of Agriculture, Food Systems, and Community Development, 5*(2), 175–199.

Pradhan, P., Lüdeke, M. K., Reusser, D. E., & Kropp, J. P. (2014). Food self-sufficiency across scales: How local can we go? *Environmental Science & Technology, 48*(16), 9463–9470.

Prange-Gstöhl, H. (2016). Eroding societal trust: A game-changer for EU policies and institutions? *Innovation: The European Journal of Social Science Research, 29*(4), 375–392.

Printezis, I., Grebitus, C., & Printezis, A. (2017). Importance of perceived "naturalness" to the success of urban farming. *Choices, 32*(1), 1–7.

Pulighe, G., & Lupia, F. (2016). Mapping spatial patterns of urban agriculture in Rome (Italy) using Google earth and web-mapping services. *Land Use Policy, 59*, 49–58.

Rahdriawan, M., & Arriani, R. R. (2020). Motives and dynamic of community-based aquaponics for urban farming in Semarang. *E&ES, 448*(1), 012096.

Rainbow, E. (2010). Experiences in development of green compost as a peat replacement material. Proceedings of The International Plants' Propagators' Society 60.

Rakocy, J. E. (2012). Chapter 14: Aquaponics – Integrating fish and plant culture. In J. H. Tidwell (Ed.), *Aquaculture production systems* (pp. 344–386). Wiley-Blackwell.

Rakocy, J. E, Bailey, D. S., Shultz, K. A. , & Cole, W. M. (1997) *Evaluation of a commercial-scale aquaponic unit for the production of tilapia and lettuce*. In: K. Fitzsimmons ed. Tilapia Aquaculture. Proceedings from the fourth international symposium on Tilapia in Aquaculture. Orlando, FL, pp. 357–372.

Rakocy, J., Shultz, R. C., Bailey, D. S., & Thoman, E. S. (2003). Aquaponic production of tilapia and basil: comparing a batch and staggered cropping system. In *South pacific soilless culture conference-SPSCC 648* (pp. 63–69).

Rakocy, J. E., Masser, M. P., & Losordo, T. M. (2006). *Recirculating aquaculture tank production systems: aquaponics – integrating fish and plant culture*. SRAC Publication No. 454. Southern Regional Aquaculture Center.

Rakocy, J. E., Bailey, D. S., Shultz, R. C., & Danaher, J. J. (2011). *A commercial-scale aquaponic system developed at the University of the Virgin Islands*. In Proceedings of the 9th international symposium on tilapia in aquaculture (pp. 336–343). AquaFish Collaborative Research Support Program.

Ramankutty, N., Foley, J. A., Norman, J., & McSweeney, K. (2002). The global distribution of cultivable lands: Current patterns and sensitivity to possible climate change. *Global Ecology and Biogeography, 11*(5), 377–392.

Ramankutty, N., Mehrabi, Z., Waha, K., Jarvis, L., Kremen, C., Herrero, M., & Rieseberg, L. H. (2018). Trends in global agricultural land use: Implications for environmental health and food security. *Annual Review of Plant Biology, 69*, 789–815.

Rao, I. M. (2009). *Essential plant nutrients and their functions*. Centro Internacional de Agricultura Tropical (CIAT). Cali, Colombia. Working Document (36).

Raymond, C. M., Giusti, M., & Barthel, S. (2018). An embodied perspective on the co-production of cultural ecosystem services: Toward embodied ecosystems. *Journal of Environmental Planning and Management*. https://doi.org/10.1080/09640568.2017.1312300

Real Food. (2017). *Grow it. Cook it. Eat it. Real Food, Wythenshawe*. Available at https://www.realfoodwythenshawe.com/wp-content/uploads/2017/11/Evaluation-Report_The-Story-so-far_Appendix-B.pdf. Accessed 30.09.2019.

Reay, D. S., Davidson, E. A., Smith, K. A., Smith, P., Melillo, J. M., Dentener, F., & Crutzen, P. J. (2012). Global agriculture and nitrous oxide emissions. *Nature Climate Change, 2*(6), 410–416.

Reisch, L., Eberle, U., & Lorek, S. (2013). Sustainable food consumption: An overview of contemporary issues and policies. *Sustainability: Science, Practice and Policy, 9*(2), 7–25.

Renkui, C., Dashu, N., & Jianguo, W. (1995). Rice–fish culture in China: The past, present, and future. In K. T. Mackay (Ed.), *Rice–fish culture in China*. International Development Research Centre.

Resh, H. M. (2013). *Hydroponic food production: A definitive guidebook for the advanced home gardener and the commercial hydroponic grower*. CRC Press.

Restrepo, J. D. C., & Morales-Pinzón, T. (2018). Urban metabolism and sustainability: Precedents, genesis and research perspectives. *Resources, Conservation and Recycling, 131*, 216–224.

Reynolds, R. (2014). *On guerrilla gardening: A handbook for gardening without boundaries*. Bloomsbury Publishing.

Ricciardi, V., Ramankutty, N., Mehrabi, Z., Jarvis, L., & Chookolingo, B. (2018). How much of the world's food do smallholders produce? *Global Food Security, 17*, 64–72.

Rigolon, A. (2016). A complex landscape of inequity in access to urban parks: A literature review. *Landscape and Urban Planning, 153*, 160–169.

Rigolon, A., Browning, M., & Jennings, V. (2018). Inequities in the quality of urban park systems: An environmental justice investigation of cities in the United States. *Landscape and Urban Planning, 178*, 156–169.

Rinker, D. L. (2002). Handling and using “spent” mushroom substrate around the world. In J. E. Sánchez, G. Huerta, & E. Montiel (Eds.), *Mushroom biology and mushroom products* (pp. 43–60). Impresos Júpiter.

Roberto, K. (2000) *How to hydroponics*. FutureGarden. ISBN - 0-9672026-0-4.

Robinson, B., Winans, K., Kendall, A., Dlott, J., & Dlott, F. (2018). A life cycle assessment of Agaricus bisporus mushroom production in the USA. *International Journal of Life Cycle Assessment, 24*, 456e467.

Rodríguez-Delfín, A., Gruda, N., Eigenbrod, C., Orsini, F., & Gianquinto, G. (2017). Soil based and simplified hydroponics rooftop gardens. In *Rooftop urban agriculture* (pp. 61–81). Springer.

Romeo, D., Vea, E. B., & Thomsen, M. (2018). Environmental impacts of urban hydroponics in Europe: A case study in Lyon. *Procedia CIRP, 69*, 540–545.

Ropohl, G. (1999). Philosophy of socio-technical systems. *Techné: Research in Philosophy and Technology, 4*(3), 186–194.

Rosmiza, M. Z., Davies, W. P., Rosniza Aznie, C. R., Jabil, M. J., & Mazdi, M. (2016). Prospects for increasing commercial mushroom production in Malaysia: Challenges and opportunities. *Mediterranean Journal of Social Sciences, 7*(1 S1), 406–406.

Rosol, M. (2012). Community volunteering as neoliberal strategy? Green space production in Berlin. *Antipode, 44*(1), 239e257.

Rotterzwam. (n.d.). *10 things you need to know when cultivating mushrooms*. Available at https://s3.amazonaws.com/kajabi-storefronts-production/sites/43424/themes/739193/downloads/OR1q1cQmR72sqJmmZmwg_Ebook1_CultivatingMushrooms_1.4.pdf. Accessed 12.06.2020.

Rubio-Portillo, E., Villamor, A., Fernandez-Gonzalez, V., Antón, J., & Sanchez-Jerez, P. (2019). Exploring changes in bacterial communities to assess the influence of fish farming on marine sediments. *Aquaculture, 506*, 459–464.

Ruggeri, G., Mazzocchi, C., & Corsi, S. (2016). Urban gardeners' motivations in a metropolitan city: The case of Milan. *Sustainability (Switzerland), 8*(11). https://doi.org/10.3390/su8111099

Russell, L. D. (2015). Democratizing the scientific space: The constellation of new epistemic strategies around the emerging metaphor of socially embedded autonomy. *Technology in Society, 40*, 82–92.

Saba, A., & Messina, F. (2003). Attitudes towards organic foods and risk/benefit perception associated with pesticides. *Food Quality and Preference*. https://doi.org/10.1016/S0950-3293(02)00188-X

Safransky, S. (2014). Greening the urban frontier: Race, property, and resettlement in Detroit. *Geoforum, 56*, 237–248.

Saha, S.K. (2010) *Soilless cultivation for landless people: An alternative livelihood practice through Indigenous hydroponic agriculture in flood-prone Bangladesh*. Beppu: Ritsumeikan Asia Pacific University. Available at https://secure.apu.ac.jp/rcaps/uploads/fckeditor/publications/journal/RJAPS_V27_Saha.pdf. Accessed: 07/08/2019.

Saha, M., & Eckelman, M. J. (2017). Growing fresh fruits and vegetables in an urban landscape: A geospatial assessment of ground level and rooftop urban agriculture potential in Boston, USA. *Landscape and Urban Planning, 165*, 130–141.

Sambo, P., Nicoletto, C., Giro, A., Pii, Y., Valentinuzzi, F., Mimmo, T., Lugli, P., Orzes, G., Mazzetto, F., Astolfi, S., & Terzano, R. (2019). Hydroponic solutions for soilless production systems: Issues and opportunities in a smart agriculture perspective. *Frontiers in Plant Science, 10*.

Sánchez, C. (2004). Modern aspects of mushroom culture technology. *Applied Microbiology and Biotechnology, 64*(6), 756–762.

Santo, R., Palmer, A., & Kim, B. (2016). *Vacant lots to vibrant plots: A review of the benefits and limitations of urban agriculture*. Johns Hopkins Center for a Livable Future. Available at https://www.researchgate.net/publication/319213554_Vacant_lots_to_vibrant_plots_A_review_of_the_benefits_and_limitations_of_urban_agriculture. Accessed 10.01.2021.

Sanyé-Mengual, E., Llorach-Masana, P., Sanjuan-Delmás, D., Oliver-Solà, J., Josa, A., Montero, J. I., & Rieradevall, J. (2014). The ICTA-ICP rooftop greenhouse lab (RTG-lab): Closing metabolic flows (energy, water, CO2) through integrated rooftop greenhouses. In *Finding spaces for productive cities* (pp. 693–701). VHL University of Applied Sciences.

Sanyé-Mengual, E., Orsini, F., Oliver-Solà, J., Rieradevall, J., Montero, J. I., & Gianquinto, G. (2015). Techniques and crops for efficient rooftop gardens in Bologna, Italy. *Agronomy for Sustainable Development, 35*(4), 1477–1488.

Sanyé-Mengual, E., Gasperi, D., Michelon, N., Orsini, F., Ponchia, G., & Gianquinto, G. (2018). Ecoefficiency assessment and food security potential of home gardening: A case study in Padua, Italy. *Sustainability, 10*(7), 2124.

Savidov, N., & Brooks, A. B. (2004). *Evaluation and development of aquaponics production and product market capabilities in Alberta*. Crop Diversification Centre South, Alberta Agriculture, Food and Rural Development.

Scheromm, P. (2015). Motivations and practices of gardeners in urban collective gardens: The case of Montpellier. *Urban Forestry & Urban Greening, 14*(3), 735–742.

Scheromm, P., & Soulard, C. T. (2018). The landscapes of professional farms in mid-sized cities, France. *Geographical Research, 56*(2), 154–166.

Schoen, V., & Blythe, C. (2020). COVID-19: A growing opportunity for community gardening in London. *RUAF blogs*. Available at https://ruaf.org/news/covid-19-a-growing-opportunity-for-community-gardening-in-london/. Accessed on 10.12.2020.

Schoen, V., Caputo, S., & Blythe, C. (2020). Valuing physical and social output: A rapid assessment of a London community garden. *Sustainability, 12*(13), 5452.

Schreefel, L., Schulte, R. P. O., de Boer, I. J. M., Schrijver, A. P., & van Zanten, H. H. E. (2020). Regenerative agriculture–the soil is the base. *Global Food Security, 26*, 100404.

Schwab, E., Caputo, S., & Hernández-García, J. (2018). Urban agriculture: Models-in-circulation from a critical transnational perspective. *Landscape and Urban Planning, 170*, 15–23.

Sennett, R. (1996). *Flesh and stone: The body and the city in Western civilization*. WW Norton.

Sharma, K., Basta, N. T., & Grewal, P. S. (2015). Soil heavy metal contamination in residential neighborhoods in post-industrial cities and its potential human exposure risk. *Urban Ecosystem, 18*(1), 115–132.

Shiklomanov, I. A., & Rodda, J. C. (2003). *World water resources at the beginning of the 21st century* (International hydrology series). Cambridge University Press.

Siegner, A., Sowerwine, J., & Acey, C. (2018). Does urban agriculture improve food security? Examining the nexus of food access and distribution of urban produced foods in the United States: A systematic review. *Sustainability, 10*(9), 2988.

Simon, H. A. (1966). Theories of decision-making in economics and behavioural science. In *Surveys of economic theory* (pp. 1–28). Palgrave Macmillan.

Singh, R. P., & Mishra, K. K. (2008). *Mushroom cultivation*.

Sismondo, S. (2004). *An introduction to science and technology studies* (Vol. 1). Wiley-Blackwell.

Social Farms & Gardens. (n.d.). *Business Heroes: Inspiring business to partner with local community growing groups*. Available at https://www.farmgarden.org.uk/system/files/business_heroes_a5_booklet_interactive.pdf. Accessed 20.0.2020.

Somerville, C., Cohen, M., Pantanella, E., Stankus, A., & Lovatelli, A. (2014). Small-scale aquaponic food production: Integrated fish and plant farming. *FAO Fisheries and Aquaculture Technical Paper, 589*, I.

Song, X. P., Tan, H. T., & Tan, P. Y. (2018). Assessment of light adequacy for vertical farming in a tropical city. *Urban Forestry & Urban Greening, 29*, 49–57.

Sonnino, R. (2009). Feeding the city: Towards a new research and planning agenda. *International Planning Studies, 14*(4), 425–435. https://doi.org/10.1080/13563471003642795

Spangler, K., Burchfield, E. K., & Schumacher, B. (2020). Past and current dynamics of US agricultural land use and policy. *Front. Sustain. Food Syst., 4*, 98. https://doi.org/10.3389/fsufs

Specht, K., & Sanyé-Mengual, E. (2017). Risks in urban rooftop agriculture: Assessing stakeholders' perceptions to ensure efficient policymaking. *Environmental Science & Policy, 69*, 13–21.

Specht, K., Siebert, R., Hartmann, I., Freisinger, U. B., Sawicka, M., Werner, A., Thomaier, S., Henckel, D., Walk, H., & Dierich, A. (2014). Urban agriculture of the future: An overview of sustainability aspects of food production in and on buildings. *Agriculture and Human Values, 31*(1), 33–51.

Specht, K., Weith, T., Swoboda, K., & Siebert, R. (2016). Socially acceptable urban agriculture businesses. *Agronomy for Sustainable Development, 36*, 1–14.

Specht, K., Zoll, F., Schümann, H., Bela, J., Kachel, J., & Robischon, M. (2019). How will we eat and produce in the cities of the future? From edible insects to vertical farming—A study on the perception and acceptability of new approaches. *Sustainability, 11*(16), 4315.

Stamets, P., & Chilton, J. S. (1983). The mushroom cultivator. *First Washington*.

Stuart, N. W. (1948). Growing plants without soil. *The Scientific Monthly, 66*(4), 273–282.

Stuart, D. (2008). The illusion of control: Industrialized agriculture, nature, and food safety. *Agriculture and Human Values, 25*(2), 177–181.

Sure Harvest. (2017). *The mushroom sustainability story: water, energy, and climate environmental metrics*. Available at https://www.mushroomcouncil.com/wp-content/uploads/2017/12/Mushroom-Sustainability-Story-2017.pdf. Accessed 10.09.2020.

Tabares, C. M. (2003). Hydroponics in Latin America. *Urban Agriculture Magazine*, p. 8. RUAF.

Taylor, J. R., & Lovell, S. T. (2012). Mapping public and private spaces of urban agriculture in Chicago through the analysis of high-resolution aerial images in Google Earth. *Landscape and Urban Planning, 108*(1), 57–70.

Theurl, M. C., Haberl, H., Erb, K. H., & Lindenthal, T. (2014). Contrasted greenhouse gas emissions from local versus long-range tomato production. *Agronomy for Sustainable Development, 34*(3), 593–602.

Thomaier, S., Specht, K., Henckel, D., Dierich, A., Siebert, R., Freisinger, U. B., & Sawicka, M. (2015). Farming in and on urban buildings: Present practice and specific novelties of zero-acreage farming (ZFarming). *Renewable Agriculture and Food Systems, 30*(1), 43–54.

Thomas, T., & Gunden, C. (2012). Investigating consumer attitudes toward food produced via three production systems: Conventional, sustainable and organic. *Journal of Food Agriculture & Environment, 10*(2), 132–135.

Thórarinsdóttir, R. I., Kledal, P. R., Skar, S. L. G., Sustaeta, F., Ragnarsdóttir, K. V., Mankasingh, U., Pantanella, E., Ven, R.V.D., & Shultz, C., (2015). *Aquaponics guidelines*.

Tilman, D., Cassman, K. G., Matson, P. A., Naylor, R., & Polasky, S. (2002). Agricultural sustainability and intensive production practices. *Nature, 418*(6898), 671–677.

Tlili, A., & Dawson, E. (2010). Mediating science and society in the EU and UK: From information-transmission to deliberative democracy? *Minerva, 48*(4), 429–461.

Todd, N., & Todd, J. (Eds.). (1976) *Journal of the New Alchemists* – Vol. 3. Available at https://newalchemists.net/publications/new-alchemy-1971-1991/. Accessed on 23.03.2020.

Tokimoto, T., Kawasaki, N., Nakamura, T., Akutagawa, J., & Tanada, S. (2005). Removal of lead ions in drinking water by coffee grounds as vegetable biomass. *Journal of Colloid and Interface Science, 281*(1), 56–61.

Tokunaga, K., Tamaru, C., Ako, H., & Leung, P. (2013). *Economics of commercial aquaponics in Hawaii*. In: Aquaponics in Hawaii conference. University of Hawaii at Manoa.

Tornaghi, C. (2014). Critical geography of urban agriculture. *Progress in Human Geography, 38*(4), 551–567. https://doi.org/10.1177/0309132513512542

Tornaghi, C. (2017). Urban agriculture in the food-disabling city:(re) defining urban food justice, reimagining a politics of empowerment. *Antipode, 49*(3), 781–801.

Tree, I. (2018). *Wilding: The return of nature to a British farm*. Pan Macmillan.

Trewavas, A. (2001). Urban myths of organic farming. *Nature, 410*(6827), 409–410.

Tricase, C., Lamonaca, E., Ingrao, C., Bacenetti, J., & Giudice, A. L. (2018). A comparative life cycle assessment between organic and conventional barley cultivation for sustainable agriculture pathways. *Journal of Cleaner Production, 172*, 3747–3759.

Tsing, A. L. (2015). *The mushroom at the end of the world: On the possibility of life in capitalist ruins*. Princeton University Press.

Turner, W. R., Nakamura, T., & Dinetti, M. (2004). Global urbanization and the separation of humans from nature. *Bioscience*. https://doi.org/10.1641/0006-3568(2004)054[0585:GUATSO]2.0.CO;2

Tziva, M., Negro, S. O., Kalfagianni, A., & Hekkert, M. P. (2020). Understanding the protein transition: The rise of plant-based meat substitutes. *Environmental Innovation and Societal Transitions, 35*, 217–223.

United Nations. (2004). *World population to 2300*. Department of Economic and Social Affairs. United Nations.

Van Brussel, S., & Huyse, H. (2019). Citizen science on speed? Realising the triple objective of scientific rigour, policy influence and deep citizen engagement in a large-scale citizen science project on ambient air quality in Antwerp. *Journal of Environmental Planning and Management, 62*(3), 534–551.

Van Woensel, L., & Archer, G. (2015). *Ten Technologies which could change our lives. Potential impacts and policy implications*. EPRS (European Parliament Research Studies). Available at https://www.europarl.europa.eu/EPRS/EPRS_IDAN_527417_ten_trends_to_change_your_life.pdf. Accessed 03.11.2019.

Varumo, L., Yaneva, R., Koppel, T., Koskela, I. M., Garcia, M. C., Sozzo, S., Morello, E., & Dictor, M. C. (2020). Perspectives on citizen engagement for the EU Post-2020 biodiversity strategy: An empirical study. *Sustainability, 12*(4), 1532.

Verhoog, H., Matze, M., Van Bueren, E. L., & Baars, T. (2003). The role of the concept of the natural (naturalness) in organic farming. *Journal of Agricultural and Environmental Ethics, 16*(1), 29–49.

Verner, D., Vellani, S., Klausen, A. L., & Tebaldi, E. (2017). *Frontier agriculture for improving refugee livelihoods: Unleashing climate-smart and water-saving agriculture technologies in MENA*. World Bank.

Viljoen, A., & Bohn, K. (2014). Urban agriculture on the map: The CPUL CITY concept. In A. Viljoen & K. Bohn (Eds.), *Second nature urban agriculture: Designing productive cities. 2004*. Routledge.

Viljoen, A., Bohn, K., & Howe, J. (2012). *Continuous productive urban landscapes*. Routledge.

Villarroel, M., Junge, R., Komives, T., König, B., Plaza, I., Bittsánszky, A., & Joly, A. (2016). Survey of aquaponics in Europe. *Water, 8*(10), 468.

Voorberg, W. H., Bekkers, V. J., & Tummers, L. G. (2015). A systematic review of co-creation and co-production: Embarking on the social innovation journey. *Public Management Review, 17*(9), 1333–1357.

Walley, K., Custance, P., & Parsons, S. (2000). UK consumer attitudes concerning environmental issues impacting the agrifood industry. *Business Strategy and the Environment, 9*(6), 355–366.

Warren, E., Hawkesworth, S., & Knai, C. (2015). Investigating the association between urban agriculture and food security, dietary diversity, and nutritional status: A systematic literature review. *Food Policy, 53*, 54–66.

Watson, C. A., Atkinson, D., Gosling, P., Jackson, L. R., & Rayns, F. W. (2002). Managing soil fertility in organic farming systems. *Soil Use and Management, 18*, 239–247.

Wheeler, R. M., Mackowiak, C. L., Sager, C. L., Knott, C. L., & Hinkle, C. R. (1990). Potato growth and yield using nutrient film technique (NFT). *Am. Potato J., 67*, 177–187.

Wheeler, A., Dykstra, P., Black, J., & Soares, N. (2020). *COVID-19: UK veg box report*. Food Foundation.

Whitworth, E., Druckman, A., & Woodward, A. (2017). Food scares: A comprehensive categorisation. *British Food Journal*. https://doi.org/10.1108/BFJ-06-2016-0263

Wielemaker, R., Oenema, O., Zeeman, G., & Weijma, J. (2019). Fertile cities: Nutrient management practices in urban agriculture. *Science of the Total Environment, 668*, 1277–1288.

Willer, H., Lernoud, J., & Kemper, L. (2018). The world of organic agriculture 2018: Summary. In *The world of organic agriculture. Statistics and emerging trends 2018* (pp. 22–31). Research Institute of Organic Agriculture FiBL and IFOAM-Organics International.

Winterborne, J. (2005). *Hydroponics: Indoor horticulture*. Pukka Press.

Wolfenson, K. D. M. (2013). *Coping with the food and agriculture challenge: Smallholders' agenda*. Food and Agriculture Organisation of the United Nations.

Woodcock, B. A., Isaac, N. J., Bullock, J. M., Roy, D. B., Garthwaite, D. G., Crowe, A., & Pywell, R. F. (2016). Impacts of neonicotinoid use on long-term population changes in wild bees in England. *Nature Communications, 7*(1), 1–8.

Yang, X., He, J., Li, C., Ma, J., Yang, Y., & Xu, J. (2008). Matsutake trade in Yunnan Province, China: An overview. *Economic Botany, 62*(3), 269–277.

Yarborough, M. (2014). Openness in science is key to keeping public trust. *Nature News, 515*(7527), 313.

Yildiz, H. Y., Radosavljevic, V., Parisi, G., & Cvetkovikj, A. (2019). Insight into risks in aquatic animal health in aquaponics. *Aquaponics Food Production Systems, 435*.

Yoo, Y. B., Oh, M. J., Oh, Y. L., Shin, P. G., Jang, K. Y., & Kong, W. S. (2016). Development trend of the mushroom industry. *Journal of Mushroom, 14*(4), 142–154.

Yoon, S. J., & Woudstra, J. (2007). Advanced horticultural techniques in Korea: The earliest documented greenhouses. *Garden History*, 68–84.

Yotti' Kingsley, J., & Townsend, M. (2006). 'Dig in' to social capital: Community gardens as mechanisms for growing urban social connectedness. *Urban Policy and Research, 24*(4), 525–537.

Zasada, I. (2011). Multifunctional peri-urban agriculture – A review of societal demands and the provision of goods and services by farming. *Land Use Policy, 28*(4), 639–648.

Zezza, A., & Tasciotti, L. (2010). Urban agriculture, poverty, and food security: Empirical evidence from a sample of developing countries. *Food Policy, 35*(4), 265–273.

Zhang, Y., Geng, W., Shen, Y., Wang, Y., & Dai, Y. C. (2014). Edible mushroom cultivation for food security and rural development in China: Bio-innovation, technological dissemination and marketing. *Sustainability, 6*(5), 2961–2973.

Websites

Aerofarms. (2021). Available at https://aerofarms.com/farms/. Accessed 19.01.2021.

Agaricus. (2020). Available at https://en.agaricus.ru/agaricus/from-the-history. Accessed 05.10.2020.

Agriculture Post. (2020). Available at https://agriculturepost.com/global-hydroponics-market-to-grow-usd-17-bn-by-2025-valuates-reports/. Accessed 13.11.2020.

Brooklyn Grange Farm. (2020). Available at https://www.brooklyngrangefarm.com/about-brooklyn-grange-1. Accessed 28.02.2020.

Circular Economy. (2019). Available at https://circulareconomy.europa.eu/platform/en/good-practices/turning-coffee-product-green-energy). Accessed 03.12.2019.

Cision. (2020). Available at https://www.prnewswire.com/news-releases/worldwide-aquaponics-industry-2020-to-2025%2D%2D-cagr-of-14-5-expected-during-this-period-301120310.html#:~:text=Owing%20to%20such%20trends%2C%20the,15.5%25%20going%20through%20to%202025. Accessed 18.11.2020.

CNG Farming. (2019). Available at https://www.cngfarming.org/aquaponics. Accessed 05.12.2019.

Community Supported Agriculture. (2021). Available at https://communitysupportedagriculture.org.uk/csa/growing-communities/. Accessed 02.02.2021.

D&B. (2020). Available at https://www.dnb.com/business-directory/top-results.html?term=thanet%20earth&page=1; and https://www.dnb.com/business-directory/company-profiles.thanet_earth_holdings_limited.7956888befb7bada1866b4b7f4c50413.html. Accessed 18.11.2020.

Debating Europe. (2020). Available at https://www.debatingeurope.eu/focus/arguments-for-and-against-the-common-agricultural-policy/#.Xr1YAhNKigA). Accessed 18.01.2020.

Edible Eastside. (2020). Available at www.edibleeastside.org. Accessed 20.02.2020.

European Commission. (2020a). Available at http://ec.europa.eu/growth/smes/business-friendly-environment/sme-definition_en. Accessed 09.06.2020.

European Commission. (2020b). Available at http://ec.europa.eu/growth/sectors/social-economy/enterprises_en). Accessed 09.06.2020.

Eurostat. (2021). *Organic farming statistics*. Available at https://ec.europa.eu/eurostat/statistics-explained/index.php?title=Organic_farming_statistics. Accessed 29.06.2021.

FAO. (2020b). Available at http://www.fao.org/faostat/en/#data/FBS/visualize. Accessed 10.04.2021.

Farm Flavor. (2019). Available at https://www.farmflavor.com/florida/walt-disney-world-farm-grows-magical-produce-earth/). Accessed 19.11.2019.

Food Assembly. (2019). Available at https://thefoodassembly.com/en-GB. Accessed 21.11.2019.

Global Generation. (2020). Available at https://www.globalgeneration.org.uk/about-the-skip-garden. Accessed 20.02.2020.

Google Trends. (2020a). Available at https://www.wordstream.com/google-trends#:~:text=Google%20Trends%20is%20a%20search,a%20given%20period%20of%20time. Accessed 03.05.2020.

Google Trends. (2020b). Available at https://www.google.com/trends. Accessed 03.05.2020.

Grand View Research. (2020b). Available at https://www.grandviewresearch.com/press-release/global-vertical-farming-market. Accessed 13.11.2020.

Growing Communities. 2020). Available at https://www.growingcommunities.org). Accessed 20.02.2020.

Hellen McArthur Foundation. (2021). Available at https://ellenmacarthurfoundation.org/topics/circular-economy-introduction/overview. Accessed 20.10.2021.

How Stuff Works. (2019). Available at https://home.howstuffworks.com/lawn-garden/professional-landscaping/hydroponics1.htm. Accessed 28.09.2019.

Incredible Edible Todmorden. (2020). Available at https://www.incredible-edible-todmorden.co.uk/). Accessed 14.02.2020.

Index of Multiple Deprivation. (2015). Available at https://opendatacommunities.org/data/societal-wellbeing/imd/indices. Accessed 21.03.2020.

Kew Gardens. (2019). Available at https://www.kew.org/kew-gardens/whats-in-the-gardens/palm-house, Accessed 28.09.2019.

LEAP. (2020). Available at https://www.madleap.co.uk/calthorpe. Accessed 07.07.2020.

Markets and Markets. (2020). Available at https://www.marketsandmarkets.com/Market-Reports/hydroponic-market-94055021.html#utm_source=PRnewswire&utm_medium=Referral&utm_campaign=PaidPR. Accessed 15.11.2020.

Mordor Intelligence. (2020). Available at https://www.mordorintelligence.com/industry-reports/aquaponics-market. Accessed 18.11.2020.

Mudlarks Garden. (2020). Available at https://mudlarksgarden.org.uk/our-team/). Accessed 15.02.2020.

New World Encyclopedia. (2019). Available at https://www.newworldencyclopedia.org/entry/Greenhouse#History. Accessed 28.09.2019.

NSALG – The National Allotment Society. (2020). Available at https://www.nsalg.org.uk/allotment-info/brief-history-of-allotments/. Accessed 14.02.2020.

Our World in Data. (2019). Available at https://ourworldindata.org/tertiary-education#all-charts-preview. Accessed 15.10.2019.

Our World in Data. (2020). Available at https://ourworldindata.org/employment-in-agriculture. Accessed 18.01.2020.

Prinzessinengarten. (2020). Available at https://prinzessinnengarten.net/about/). Accessed 20.02.2020.

Prospect Farm. (2019). Available at https://prospectfarm.org/2017/05/08/experiments-in-urban-farming/). Accessed 03.12.2019.

Real Food Wythenshawe. (2020). Available at https://www.realfoodwythenshawe.com/about/). Accessed 21.02.2020.

Roof Water Farm. (2020). Available at http://www.roofwaterfarm.com/en/ueber/). Accessed 15.03.2020.

Skyfarn. (2021). Available at http://skyfarm.tech/post-01/. Accessed 02.02.2021.

Social Farms & Gardens. (2021). Available at Social Farms and Gardens, 2020. https://www.farmgarden.org.uk/news/crop-cycle-welsh-government-fund-community-benefits-across-4-sites-wales. Accessed 10.01.2021.

Statista. (2020a). Available at https://www.statista.com/statistics/1226211/population-of-africa-by-age-group/. Accessed 03.12.2021.

Statista. (2020b). Available at https://www.statista.com/statistics/1072288/agricultural-planted-area-brazil/. Accessed 01.12.2020.

Sydenham Garden. (2020). Available at https://www.sydenhamgarden.org.uk/project/growing-lives/. Accessed 15.02.2020.

The GCRI Trust. (2019). Available at http://www.gcritrust.org/history-gcri/. Accessed 15.10.2019.

Two Wests. (2019). Available at https://www.twowests.co.uk/blog/the-history-of-hydroponics). Accessed 13.10.2019.

Urbed. (2020). Available at http://urbed.coop/vertical-farm-manchester. Accessed 21.03.2020.

Veniverdi. (2020). Available at https://www.veniverdi.fr/. Accessed 20.02.2020.

Verdes del Sur. (2020). Available at http://huertosverdesdelsur.blogspot.com/p/el-milagro-de-los-peces.html. Accessed 04.07.2020.

Verti-gro. (2019). Available at www.vertigro.com. Accessed 28.11.2019.

World Bank. (2020). Available at https://data.worldbank.org/indicator/NV.AGR.TOTL.ZS?end=2018&start=1960. Accessed 18.01.2020.

Zipgrow. (2019). Available at www.zipgrow.com. Accessed 28.11.2019.

Index

S. Caputo, *Small Scale Soil-less Urban Agriculture in Europe*, Urban Agriculture, https://doi.org/10.1007/978-3-030-99962-9

The manufacturer's authorised representative in the EU is Springer Nature Customer Service Centre GmbH, Europaplatz 3, 69115 Heidelberg, Germany. If you have any concerns regarding our products, please contact ProductSafety@springernature.com

Printed and bound by CPI Group (UK) Ltd, Croydon, CR0 4YY

01/10/2026

02225872-0002